William Henry Spencer

Elements of qualitative chemical analysis

William Henry Spencer

Elements of qualitative chemical analysis

ISBN/EAN: 9783337279868

Printed in Europe, USA, Canada, Australia, Japan

Cover: Foto ©berggeist007 / pixelio.de

More available books at **www.hansebooks.com**

ELEMENTS

OF

TIVE CHEMICAL ANALYSIS.

BY

W. H. SPENCER, B.A.

FELLOW OF THE CAMBRIDGE PHILOSOPHICAL SOCIETY.

London and Cambridge:

MACMILLAN AND CO.

1866.

PREFACE.

In the following compilation the chief reactions of the more important basic and acid radicals are so correlated and contrasted, as to exhibit in a brief and convenient manner the principles upon which various methods of analysis are founded; and a frequent error is avoided—that of combining the Practical Manual for the Laboratory with the Descriptive Treatise on Elementary Chemistry.

The reactions are tabulated in the First and Second Parts. The various methods which constitute the systematic course of Analysis are detailed and arranged, with special reference to the convenience of the working student, in the Third and Fourth Parts.

This work is designed to serve as a Manual for the beginner, and as a Text-book for examination; its original character consists in the *arrangement* of information which may be found in existing Manuals of Chemical Analysis.

In an Introduction will be found details as to the mode in which the work should be used, based upon the experience gained in teaching, by this method, the Elements of Analysis.

W. H. S.

January, 1866.

CONTENTS.

PAGE

Introduction . . . 1—9
Illustration of the method of analysis . . . 6
Table of atomic weights . . . 8
Classification of Basic radicals . . . 9

PART I. Basic Radicals.

Blowpipe Manipulation . . . 12
Behaviour of Basic Radicals with Reagents in the Dry Way . . . 13—19
Group I. . . . Salts of Silver . . . 13
" Lead . . . 13
" Mercury . . . 13
Group II. Sect. I. . . . " Tin . . . 14
" Antimony . . . 14
" Arsenic . . . 14
" Platinum . . . 14
" Gold . . . 14
Group II. Sect. II. . . . " Bismuth . . . 15
" Cadmium . . . 15
" Copper . . . 15
Group III. . . . " Aluminium . . . 16
" Chromium . . . 16
" Iron . . . 16
Group IV. . . . " Manganese . . . 17
" Zinc . . . 17
" Cobalt . . . 17
" Nickel . . . 17
Group V. . . . " Barium . . . 18
" Strontium . . . 18
" Calcium . . . 18
" Magnesium . . . 18
Group VI. . . . " Potassium . . . 19
" Sodium . . . 19
" Ammonium . . . 19
" Lithium . . . 19

The Process of Solution . . . 20
Behaviour of Basic Radicals with Reagents in the Wet Way . . . 22—39
Group I. . . . Salts of Silver . . . 22
" Lead . . . 22
" Mercury (mercurous) . . . 22
Group II. Sect. I. . . . " Tin (stannous) . . . 24
" " (stannic) . . . 24
" Antimony (antimonious) . . . 24
" " (antimonic) . . . 25
" Arsenic (arsenious) . . . 25
" " (arsenic) . . . 25
" Platinum . . . 26
" Gold . . . 26

Special Tests for Arsenic . . . .
Marsh's Test . . . .
Reinsch's Test . . . .
Fresenius' and Babo's Test . . . .
Group II. Sect. ii. . Salts of Mercury (mercuric) . . . .
" Bismuth . . . . . .
" Cadmium . . . . . .
" Copper . . . .
Group III. . . . . " Aluminium . . . . .
" Chromium . . . . . .
" Iron (ferric) . . . . .
" " (ferrous) . . . .
Group IV. . . . . " Manganese . . . . .
" Zinc . . . . . .
" Cobalt . . . .
" Nickel . . . . . .
Group V. . . . " Barium . . . .
" Strontium . . . . . .
" Calcium . . . . . .
" Magnesium . . . .
Group VI. . . " Potassium . . . . .
" Sodium . . . . . .
" Ammonium . . . . .
" Lithium . . . .

PART II. Acid Radicals.

Classification of Acid Radicals and their Hydrogen Salts . . .
Behaviour of Inorganic Acid Radicals with Reagents . . .
Group I. Sect. i. . . Salts of the Sulphuric radical. Sulphates .
" Silicofluorine. Fluosilicates .
Sect. ii. . . " the Carbonic radical. Carbonates .
" Silicic " Silicates . .
Sect. iii. . . " Phosphoric " Phosphates .
" Boracic " Borates .
" Oxalic " Oxalates .
" Fluorine. Fluorides .
Sect. iv. . . " Chromium. Chromates .
" the Sulphurous radical. Sulphites .
" Hyposulphurous " Hyposulphites
" Iodic " Iodates .
Group II. Sect. i. . . " Chlorine. Chlorides .
" Bromine. Bromides .
" Cyanogen. Cyanides .
" Ferri-cyanogen. Ferricyanides
Sect. ii. . . " Iodine. Iodides . .
" Sulphur. Sulphides .
" Ferro-cyanogen. Ferrocyanides .
" Sulpho-cyanogen. Sulphocyanides
Group III. . . . . " the Nitric radical. Nitrates . .
" Chloric " Chlorates .
Behaviour of Organic Acid Radicals with Reagents . . . .
Group I. . . Salts of the Tartaric radical. Tartrates .
" Citric " Citrates . .
Group II. . . . . " Succinic " Succinates .
" Benzoic " Benzoates .
" Gallic " Gallates .
" Tannic " Tannates .
Group III. . . . . " Acetic " Acetates .
" Uric " Urates . .

PART III. Basic Radicals.

The Systematic Course of Analysis.

	PAGE
Introduction	60
Preliminary Examination	62
Behaviour of Basic Radicals and certain Acid Radicals with Group reagents	64
Outline of the Systematic Course of Analysis	66
The Systematic Course of Analysis for Basic Radicals	68–78
Method of Analysis for Group I.	68
Treatment of an acid or a neutral solution with Hydrochloric acid	68
Examination for Silver, Lead, and Mercury	68
Treatment of an alkaline solution with Hydrochloric acid	69
Method of Analysis for Group II.	70
Treatment of an acid solution with Hydrosulphuric acid	70
Examination for Platinum and Gold	70
,, ,, Mercury, Lead, Bismuth, Cadmium and Copper	71
Hofmann's method for separation of Cadmium and Copper	72
Examination for Arsenic, Antimony and Tin	72
Hofmann's method for separation of Arsenic, Antimony and Tin	73
Method of Analysis for Group III.	74
Treatment of a solution with Hydrate of Ammonium	74
Examination for Chromium, Aluminium and Iron	75
Method of Analysis for Group IV.	76
Treatment of a solution with Sulphide of Ammonium	76
Examination for Manganese, Zinc, Cobalt and Nickel	76
Method of Analysis for Group V.	77
Treatment of a solution with Carbonate of Ammonium	77
Examination for Barium, Strontium and Calcium	77
Method of Analysis for Group VI.	78
Examination for Ammonium, Magnesium, Potassium and Sodium	78

PART IV. Acid Radicals.

The Systematic Course of Analysis.

Introduction	80
Preliminary Examination	82
Preparation of the Solution	83
Behaviour of Acid Radicals and their Salts with General reagents	84
The Systematic Course of Analysis for Acid Radicals	86–90
Method of Analysis for Group I. (Inorganic and organic acids)	86
Treatment of a neutral solution with Chloride of Barium	86
Examination for Sulphuric acid and Hydrofluosilicic acid	86
,, ,, Phosphoric acid, Oxalic acid, Hydrofluoric acid, Silicic acid, Boracic acid, Citric acid and Tartaric acid	87
Method of Analysis for Group II.	88
Treatment of an acid solution with Nitrate of Silver	88
Examination for Hydriodic acid, (Iodic acid,) Hydrosulphocyanic acid and Hydroferrocyanic acid	88
Examination for Hydrobromic acid, Hydrochloric acid, Hydrocyanic acid and Hydroferricyanic acid	89

	PAGE
METHOD OF ANALYSIS FOR GROUP III.	90
Examination for Nitric acid and Chloric acid	90
METHOD OF ANALYSIS FOR GROUP II. (Organic acids)	90
Treatment of a neutral solution with Ferric Chloride	90
Examination for Succinic acid, Benzoic acid, Gallic acid, Tannic acid and Acetic acid	90
DETECTION OF ACIDS NOT INCLUDED IN THE SYSTEMATIC COURSE—viz. Carbonic acid, Hydrosulphuric acid, (Hydrocyanic acid,) Sulphurous acid and Hyposulphurous acid	90

APPENDIX.

ANALYSIS OF SUBSTANCES INSOLUBLE IN WATER AND IN ACIDS	91, 93
ANALYSIS OF ALKALINE SILICATES	93
ANALYSIS OF ALLOYS	94

ERRATA.

PAGE.	ERROR.	CORRECTION.
6, line 4 (from bottom).	p. 11	p. 13
13, Mercury, note 1.	tribe	tube
,, ,, ,,	expelled by gentle heat	gently heated
18, Barium, note 1.	chlorides, bromides and iodides	chloride, bromide and iodide
,, Strontium, note 1.	,, ,, ,,	,, ,, ,,
,, Calcium, note 1.	,, ,, ,,	,, ,, ,,
,, ,, note 2.	oxide or undecomposed...and iodide	oxide and undecomposed...or iodide
24, Tin (stannous salts), I.	SnCl	2SnCl
,, Tin (stannic salts), note α.	stannous chloride	stannic chloride
25, Antimony (antimonic salts), I.	$2K_2SbO_3$	$2KSbO_3$
,, Arsenic (arsenic salts), note γ.	more	only
29, Experiment, last paragraph.	and after dissolving in water may be further	the mirror dissolved in water and
30, Bismuth, IV.	cystalline	crystalline
31, Copper, note δ.	1 eq of hydrogen	1 atom of hydyrogen
36, Strontium, note γ.	also throws down	also throw down
45, Silicic acid, note ε.	absolutely soluble	absolutely insoluble
46, Phosphoric acid, IV.	$(Fe_2)_3PO_4$	Fe_2PO_4
,, ,, ,, note α.	$P_2O_5+2H_2O=3H_3PO_4$	$P_2O_5+3H_2O=2H_3PO_4$
47, Hydrofluoric acid, VI. i.	fluoride	fluoride. β
49, Hyposulphurous acid, IV. line 2.	Section 1.	Section IV.
,, ,, ,, note β.	sulphite	hyposulphite
54, Nitric acid, heading to column.	nitric acid (HNO)	nitric acid (HNO_3) a.
,, II. (throughout)	sulpindigotic acid	sulphindigotic acid
57, Tannic acid, I.	$H_2C_{27}H_{19}O$	$H_3C_{27}H_{19}O_{17}$
62, col. III. line 4.	sulphate	sulphite
64, Hydrosulphuric acid, III.	3.	4.
83, line 9.	hydrochloric acid might	hydrochloric acid in excess might
86, line 13.	alkaline	ammoniacal
,, line 14.	hydrate of ammonium	hydrate of potassium
87, col. I. line 3.	Oxalic acid	Hydrofluoric acid
,, col. I. line 4.	Hydrofluoric acid	Oxalic acid
,, col. III. line 14.	$(Fe_2)_3PO_4$	Fe_2PO_4
88, col. II. line 8 (from bottom).	50	52
89, col. III. line 2 (from bottom).	p. 50	p. 51

N.B. The corrections should be made *before* the work is used.

INTRODUCTION.

CONCERNING every material substance there are three things which it is the province of Chemical Science, theoretical or practical, to ascertain :—

First—The *quality* of its constituents. Simply what they are. In the compound, chloride of potassium (KCl), the basic radical Potassium and the acid radical Chlorine are associated as the ultimate elements of which the body is composed. The appropriate application of certain tests detects the quality of the constituents of any compound.

Second—The *quantity* of its constituents. The relation of their proportion. In the foregoing example, the quantity of Potassium in grains, or in terms of any standard we may select, combined with a certain quantity of Chlorine, calculated in terms of the same standard. The application of the balance reveals the quantity of the constituents of any compound.

Third—The *order of arrangement* of its constituents. This is a matter of far greater difficulty than to determine the quality or quantity of the bodies which make up any given substance. It involves questions indicated by such terms as 'formulæ,' 'atomic weight,' &c., and other considerations of a purely theoretical character.

Analytical Chemistry deals with the first and second of the above aspects of bodies, but at the same time, in making manifest the conclusions deduced from observation and experience, avails itself of the third question, and introduces theoretical matters in making known its results.

The whole scheme of operations by which is ascertained the *quality* of the constituents of any substance, and subordinately the properties of any particular body which serve to distinguish it from other bodies, is termed *Qualitative Analysis.*

Proportion and quantity, in reference to any combination, is the subject of another branch of analytical operations, viz. *Quantitative Analysis.*

To *identify* a body, is, in chemical language, to *test* it; that is, to apply such agents in a particular manner, as shall most easily and appropriately bring into view certain characteristic properties or some special property of that body. This is the end of ordinary Qualitative Analysis as usually practised by the student, and is attained with comparatively little difficulty.

To *separate* the various constituents of any body or mixture of substances, is liable to be a more serious matter, and forms as it were the link between mere identification and the quantitative estimation of substances when mixed or chemically combined.

In the present work we are chiefly concerned to ascertain *what* properties, singly or collectively, exhibited by those bodies, more often met with by the analyst, when brought in presence of certain other bodies, may be taken as a means of identifying or separating them.

The deportment of chemical substances, under certain conditions and when brought in contact with other particular chemical substances—the behaviour of bodies with reagents—in other words, 'reactions,' form an extensive and complicated basis upon which the several modes of analysis are founded. It is clear that these reactions must be fully comprehended and remembered before such details can be applied to any method of analysis. To facilitate the application of reagents to the detection and separation of the numerous bodies with which the analyst is called upon to deal, and to aid the memory in remembering the action of such reagents, a certain conventional classification is adopted. In the first place, bodies are divided into Basic and Acid radicals, simple or compound. Potassium is a simple basic radical and Ammonium a compound one. Similarly Chlorine represents a simple, and Cyanogen (CN) a compound, acid radical. Each of these primary classes is again subdivided into *Groups*, having reference to the behaviour of each member of any particular group with some especially selected reagent, hence called the Group-test: thus Silver, Lead, and Mercury in the form of its mercurous salts, all form with hydrochloric acid, insoluble chlorides. The chlorides of almost all other basic radicals are soluble. Hence we obtain a ready method of separating any or all of the metals of Group I. from solutions which may contain other metals. It is true that under particular conditions, the application of reagents with a view to separate groups, is not so simple as this mode of representing it may appear; but the exceptional difficulties which arise constitute so many precautions which have to be exercised by the careful analyst. And here again, an intimate knowledge of reactions is absolutely necessary to a successful prosecution of any analytical scheme. By taking advantage of certain characters pertaining only to a limited number of members in some of the larger groups, these become again subdivided into smaller assemblages; in this way Group II. (Basic radicals) is split up into two sections, depending on the solubility of the sulphides of Section I., and the insolubility of those of Section II. in sulphide of ammonium. By the application of less general tests, a still further process of division and isolation is effected: thus, it will be observed that chloride of silver is insoluble in water, mercurous chloride is also insoluble, whilst chloride of lead dissolves in boiling water. This series of facts enables the separation of Lead to be readily effected from the precipitate thrown down by the group-test. The reactions again of argentic and mercurous chloride with hydrate of ammonium furnish a method of separating Silver from Mercury. Thus the exhaustive process is carried to its extreme limit, and several basic radicals previously contained in the same solution, it may be forming different salts, and in this state undistinguishable from one another, are isolated and obtained in a condition to be recognised by means of a final confirmatory test. The selection of this most special test is to a great extent a matter of choice on the part of the operator, provided only that it is conclusive as to the presence or absence of the particular radical whose recognition or otherwise he has in view. A careful comparison of the several steps in the appended "Method of Analysis for Group I. of the Basic radicals," with the reactions to which they refer, will not only fully illustrate the mode in which reactions are applied in actual analysis, but indicate the value of the method of arrangement adopted in this work.

In attempting to apply a similar set of processes to the detection and separation of the acid radicals greater difficulty will be experienced. From the peculiar nature of their reactions, it is almost impossible to obtain a sufficient number of characters, graduating from the general to the special, to enable the analyst to proceed in a manner so regularly exhaustive: greater dependence must therefore be placed upon the repeated application of more special tests, and especially upon the results obtained on submitting the substance to be analysed, to certain processes of decomposition, e.g. the decompositions effected by sulphuric acid.

So far it has been assumed that the substance with which the analyst is concerned, is in the liquid form. But this is by no means usually the case—generally the operator at the outset of his investigations has to do with matter in the solid condition, and the process of solution is effected only after certain results have been obtained by a series of experiments, or reactions, performed in "the dry way." The first set of tables refer to the behaviour of basic radicals with reagents in the dry way. The information elicited on exposing a solid body to a high temperature, either alone, or in presence of other bodies generally also in the solid form, is not so definite or reliable as that obtained on treating the body when in solution, in what is technically known as "the wet way." Nevertheless, the performance of the blowpipe experiments should never be omitted, since much valuable light is frequently thrown upon the subsequent method of analysis by the results of this "preliminary examination." The student should, however, never allow any information so furnished to induce him to alter or abbreviate the systematic course of analysis in the wet way. Considerable experience and skill on the part of the operator can alone render such attempts successful. The whole course of systematic qualitative analysis, as thus marked out, may be divided into three parts.

I. Preliminary Examination (the application of reagents in the dry way).

II. Solution (the process of converting a solid body into the liquid form).

III. Analysis in the wet way (the application of reagents in solution).

The subordinate operations included in each of these primary divisions are susceptible of a considerable amount of variation, all, however, conducing to the same end. It will be clear that an orderly and systematic course must be followed in any method of analysis, to whatever extent the details of systems may differ. This work, in its arrangement, seeks to induce a methodical acquaintance with the details of Analysis. A few words as to the manner in which the work should be studied.

A knowledge of the reactions, both of basic and acid radicals (as detailed in Parts I. and II.), *is essential, before proceeding to the course of actual analysis.*

The radicals are classed in groups, and the reactions of each group are arranged to form a separate table.

As the value of each table, in relation to the analytical course, depends on the manner in which it is used, it will be advisable to explain the construction of the tables for some one group, selected as an example, e.g. the tables for Group IV. pp. 17 and 34.

Under the name of each radical (Manganese, &c.) is placed the name of the salt recommended to be employed in the experiments (sulphate of manganese, &c.).

Table on page 17. Heat sulphate of manganese alone, in the outer blowpipe-flame; moisten the residue so obtained with nitrate of cobalt, and heat again. Perform the same experiment with sulphate of zinc, and so on with each salt in succession. Next heat each salt separately in the same order, alone, in the inner flame. Then proceed to fuse each salt in the same order with borax on platinum wire, and so on through the experiments.

Table on page 34. Add sulphide of ammonium to sulphate of manganese; try the solubility or insolubility of the resulting flesh-coloured sulphide of manganese (Mn_2S) in acids, as mentioned. Next add sulphide of ammonium to sulphate of zinc, and so on for each radical. Then proceed to the second series of experiments with hydrate of potassium and sulphate of manganese, sulphate of zinc, &c. in succession.

Thus, the order of the experiments and the reagents to be employed, are indicated in the extreme left-hand *vertical* columns; the similarity or differences in the reactions of the various radicals with the same reagent, are exhibited in the *horizontal* columns; the reactions of any particular radical with the selected reagents, are contained in the *vertical* column under that radical.

Each group should be taken in the order in which it occurs, and completed before proceeding to the next. In the case of any group, the experiments relating to reactions in the *dry way* are to be performed first, then those relating to reactions *with solutions*, for the same group.

The notes relating to any particular reaction should be carefully studied *before* performing the experiments. It is intended that the student should himself supply the formulæ required to complete the equations, the latter being completed only in particular instances.

It will be evident that many of the reactions detailed in the First and Second Parts of this work have little or no reference to the methods by which the radicals themselves are sought for in actual analysis. But each reaction will afford, under appropriate circumstances, a mode of recognising the radical to which it refers, and the careful performance of each experiment will do more to impress upon the learner the characteristics of the numerous salts thus brought under his observation, than any amount of book-reading.

Having mastered the reactions in the tables (both in the dry way and in solution) for any one group, the student should then *construct for himself a method for the detection of one, or separation of two or more of its members, by the wet way*, and apply such a plan to the actual analysis of a salt containing one member of the group, *previously making a preliminary examination* based upon the reactions in the dry way. The analysis of a series of salts should then be undertaken, each salt containing a different member of the group, until certainty and facility are obtained in the determination of any of its members.

The next group may then be taken and a similar plan adopted, and so on for each group.

The Tables of reactions for acid radicals should be studied in a similar manner.

The next step is to determine both the basic and acid constituent in the same single salt (this however may be omitted at discretion), and, finally, the course of analysis, made as complicated as the student's powers will allow, may be undertaken according to the methods detailed in Parts III. and IV.

Whenever the analysis of any substance is commenced, with a view to determine the basic or the acid constituent, or both, every step should be carefully written out, as it is taken. It will be found convenient to record these notes in three parallel columns; the experiment performed, in the first, the result of the experiment, in the second, and the inference as to the presence or absence of any bodies, in the third. Thus:

Experiment.	Result.	Inference.
Hydrochloric acid, in an acid solution.	White precipitate.	Presence of Silver, Lead, or Mercury.
Filter.		
Precipitate, washed with cold water.		
And boiled with water.	Precipitate entirely dissolves.	Presence of Lead.
Sulphuric acid, added to solution in water.	White precipitate.	Confirms presence of Lead.

The plan of recording notes of analysis will, however, require modification according to circumstances. At an examination, the above plan will be found to possess advantages. (See also next page).

METHOD OF ANALYSIS FOR GROUP I.

The analysis of the members of this group is commenced by converting them into chlorides. The reactions of their chlorides with water and hydrate of ammonium (Gr. I. col. I.)* afford the necessary conditions for detection or separation.

Group-test = Hydrochloric Acid (Gr. I. col. I.)

The solution to be examined is either acid or neutral.

(A few drops of the acid will in this case be sufficient to determine the presence or absence of members of this group.)

Hydrochloric acid fails to precipitate Lead from *dilute* solutions (Gr. I. col. I. note *a*); hence the non-formation of a ppt. upon adding hydrochloric acid does not prove the absence of this metal. If present it will be detected in the analysis of the following group. (Gr. I. col. II.)

Hydrochloric acid precipitates along with members of this group, *Oxychlorides of Antimony and Bismuth*, from their solutions; these re-dissolve in an excess of the precipitant. (Gr. II. sect. I. col. III. Gr. II. sect. II. col. III.)

Analysis of Precipitate produced by Hydrochloric Acid, in an Acid or Neutral Solution.

Filter.	Filtrate may contain members of other groups.	Set aside for further examination.
Wash the ppt. with cold water and add the washings to the filtrate.	Turbidity on adding the washings.	Presence of Antimony or Bismuth (Gr. II. Sect. I. and II. col. III.).
Treat the ppt. with boiling water (Gr. I. col. I.). Filter.	Filtrate may contain Lead.	Add sulphuric acid. white ppt. = *Sulphate of Lead* (Gr. I. col. VII.).
Treat the ppt. with hydrate of ammonium (Gr. I. col. I.). Filter.	Filtrate may contain Silver.	Add nitric acid (Gr. I. col. I.). white ppt. = *Nitrate of Silver*.
Black residue (Gr. I. col. I.).	Indicates presence of Mercury.	Dry the residue. Mix with carbonate of sodium. Heat in a bulb-tube (p. 11, Gr. I. col. II.). Grey sublimate = *Metallic Mercury*.

* The columns referred to are in all cases the *horizontal* columns.

Precipitate produced by Hydrochloric Acid in an Alkaline Solution.

The formation of this precipitate depends on the fact that certain salts, not only of this group, *dissolve in alkalis or alkaline salts,* which do not dissolve in water or hydrochloric acid. On adding hydrochloric acid, the alkali or alkaline salt is decomposed, and the salt held in solution separates.

Thus chloride of silver dissolves in hydrate of ammonium (Gr. I. col. I.);

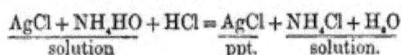

$$\underbrace{AgCl + NH_4HO}_{\text{solution}} + HCl = \underbrace{AgCl}_{\text{ppt.}} + \underbrace{NH_4Cl + H_2O}_{\text{solution.}}$$

Tersulphide of antimony dissolves in sulphide of ammonium (Gr. II. sec. I. col. I.);

$$\underbrace{Sb_2S_3 + (NH_4)_2S}_{\text{solution}} + 2HCl = \underbrace{Sb_2S_3}_{\text{ppt.}} + \underbrace{2NH_4Cl}_{\text{solution}} + \underbrace{H_2S}_{\text{escapes as gas.}}$$

In a similar manner if an alkaline carbonate or cyanide is the cause of solution, on adding hydrochloric acid, decomposition ensues, with evolution of carbonic anhydride or hydrocyanic acid gas.

Further details as to the contingencies which might arise from the addition of hydrochloric acid to an alkaline solution, are omitted here, since from their complexity they would be liable to lead to confusion, and are not of any practical value to the beginner.

TABLE OF ATOMIC WEIGHTS

EMPLOYED IN THE FORMULÆ.

Basic elements—			*Acid elements—*			
Potassium	K.	39·	Chlorine	Cl.	35·5	
Sodium	Na.	23·	Bromine	Br.	80·	
Lithium	Li.	7·	Iodine	I.	126·	
Barium	Ba.	68·5	Fluorine	Fl.	19·	
Strontium	Sr.	44·	*Oxygen*	O.	16·	(doubled)
Calcium	Ca.	20·	*Sulphur*	S.	32·	(doubled)
Magnesium	Mg.	12·	*Carbon*	C.	12·	(doubled)
Aluminium	Al.	13·7	Boron	Bo.	11·	
Chromium	Cr.	26·	Silicon	Si.	22·	
Iron	Fe.	28·	Nitrogen	N.	14·	
Manganese	Mn.	27·5	Phosphorus	P.	31·	
Cobalt	Co.	30·	Arsenic	As.	75·	
Nickel	Ni.	29·				
Zinc	Zn.	32·6				
Cadmium	Cd.	56·				
Copper	Cu.	32·				
Silver	Ag.	108·				
Mercury	Hg.	100·				
Lead	Pb.	103·5				
Bismuth	Bi.	208·				
Tin	Sn.	58·				
Antimony	Sb.	120·3				
Platinum	Pt.	98·7				
Gold	Au.	197·				
Hydrogen	H.	1·				

CLASSIFICATION OF THE BASIC RADICALS.

Group I.

Basic radicals, which are precipitated from their solutions as Chlorides, by *Hydrochloric Acid.*

Silver.
Lead.
Mercury (mercurous salts).

Group II.

Basic radicals, which are precipitated from their acidified solutions as Sulphides, by *Hydrosulphuric Acid.*

Section I.	Section II.
Sulphides, soluble in *Sulphide of Ammonium.*	Sulphides, insoluble in *Sulphide of Ammonium.*
Tin.	Mercury (mercuric salts).
Antimony.	Bismuth.
Arsenic.	Cadmium.
Platinum.	Copper.
Gold.	

Group III.

Basic radicals, which are precipitated from their solutions as Hydrates, by *Hydrate of Ammonium in the presence of Chloride of Ammonium.*

Aluminium.
Chromium.
Iron (ferric salts).

Group IV.

Basic radicals, which are precipitated from their solutions as Sulphides, by *Sulphide of Ammonium.*

Manganese.	Nickel.
Zinc.	Iron (ferrous salts).
Cobalt.	

Group V.

Basic radicals, which are precipitated from their solutions as Carbonates, by *Carbonate of Ammonium.*

Barium.	Calcium.
Strontium.	Magnesium.

Group VI.

Basic radicals, which are *not* precipitated from their solutions by *reagents which precipitate the other groups.*

Potassium.	Ammonium.
Sodium.	Lithium.

PART I.

BASIC RADICALS.

THE BLOWPIPE FLAME consists of an *outer* and an *inner* part.

The *outer* or *oxidising* flame is the outer yellow cone.

The greatest oxidising effect is produced *at* or *just beyond* the point of this cone, since the substance to be oxidised is here strongly heated and at the same time is in immediate contact with the oxygen of the air.

The *inner* or *reducing* flame is the inner blue cone.

The greatest reducing effect is produced *just beyond* the point of this cone, since the substance to be reduced is here subjected to the greatest heat, and at the same time brought in contact with unburned matter of the flame, ready to take up oxygen.

All substances should be reduced to a fine state of subdivision in a mortar, if not already in that condition, before subjecting them to the action of the blowpipe flame.

The flame should be allowed to impinge *slowly* on the charcoal or other support.

To form a BORAX BEAD on platinum wire, the wire should be twisted at one end into a loop, the moistened wire then dipped into Borax and heated in the outer blowpipe flame, if necessary more Borax may be taken up, and again heated, and so on until a clear, round bead is formed on the loop of the wire.

The smallest possible quantity of the substance to be experimented upon, should be allowed to adhere to the Borax bead.

FLUXES (such as carbonate of sodium) should be reduced to a fine state of subdivision, and *intimately mixed* with the substance to be fused, before applying heat.

N.B. The conventional terminations *-ous* and *-ic*, frequently employed throughout these pages, have reference to the relative proportions in any salt, of its basic and acid constituents.

The salt which contains
the greatest number of equivalents of basic radical
to the least number of equivalents of acid radical } = *-ous* salt, e.g. Ferrous Sulphate $Fe_2 (So_4)$.

The salt which contains
the least number of equivalents of basic radical
to the greatest number of equivalents of acid radical } = *-ic* salt, e. g. Ferric Chloride $Fe_2 Cl_6$.

BEHAVIOUR OF BASIC RADICALS WITH REAGENTS IN THE DRY WAY.

N.B. The Blowpipe is to be used in all the following experiments, unless the contrary is stated.

GROUP I.

	SILVER Nitrate of Silver.	**LEAD** Acetate of Lead.	**MERCURY** Mercurous Salts Subnitrate of Mercury. Mercuric Salts Chloride of Mercury.
Heated on Charcoal. i. In the *outer* flame.	non-volatile. Dark-red incrustation[1].	non-volatile. Coloured incrustation. Orange (hot). Yellow (cold).	volatile. Dense white incrustation.
ii. In the *inner* flame with Na_2CO_3.	White metallic globules[2]. *malleable.*	Metallic globules. *malleable*[1]. With yellow incrustation[2].	
Fused with Carbonate of Sodium. In a glass bulb-tube, or tube closed at one end.			Grey sublimate, on cool part of the tube[1].
	[1] This does not occur with all argentic salts. [2] The best way to render reduced metallic grains visible, as well as to test their malleability, in all cases, is to scrape off the reduced mass from the charcoal, and wash it by triturating in a small mortar with distilled water, when the metal becomes distinctly visible.	[1] The metal may be easily reduced without the aid of carbonate of sodium. [2] This incrustation is produced, because metallic lead is volatile at a red heat, and the oxide itself at a still higher temperature. The *outer flame* in this experiment acquires a blue colour, distinguishing between lead and bismuth.	[1] The salt is decomposed, the metal volatilizes and condenses upon the cooler neck of the tube. $2HgCl + Na_2CO_3 = Hg_2CO_3 + 2NaCl$ and $Hg_2CO_3 = \underset{\text{volatile}}{2Hg} + CO_2 + O$ The mercury salt must be anhydrous, and therefore should be dried, and the carbonate of sodium ignited, before use: with certain salts however (chlorides) which are volatile without decomposition, the mixture must be moistened with water, and then expelled by gentle heat, before applying the blowpipe flame.

GROUP II. Section i.

	TIN. Stannous Salts Protochloride of Tin. Stannic Salts Bichloride of Tin.	**ANTIMONY.** Antimoniate of Potassium.	**ARSENIC.** Arsenious Acid.	**PLATINUM.** Bichloride of Platinum.	**GOLD.** Terchloride of Gold.
Heated on Charcoal. i. In the *outer* flame. α. Alone.	Stannic salts. **Bluish-white incrustation.** (distant). Stannous salts. **Dense white incrustation.**	**White incrustation.** (distant).	Volatile. **White incrustation.** (distant.) **Odour of garlic.**	**Black metallic mass or powder**[1].	
β. Residue moistened with $CoNO_3$.	**Bluish-green colour.**	——	——	——	——
ii. In the *inner* flame. With Na_2CO_3+KCy.	**Metallic globules.** *malleable*[1].	**Metallic globules.** *brittle.* **With white incrustation**[1], *outer flame coloured bluish-green.*	——	——	**Yellow metallic globules.** *malleable.*
Heated in a Glass Tube. i. Open at both ends and held obliquely in ordinary flame.	——	**White spicular sublimate**[2].	**White crystalline sublimate**[1] in cool part of tube. **Garlic odour.**	——	——
ii. Closed at one end and heated in blowpipe flame. α. Alone.			**Mostly volatilize unchanged**[2]. **Black deposit** (if the substance is *uncombined Arsenic*). **Yellow deposit** (if a *Sulphide*). **Red crystalline deposit** (if an *Iodide*).		
β. With Na_2CO_3 and Charcoal[1].	——	——	**Metallic mirror** on upper part of tube.	——	——
[1] The arsenical compound must be intimately mixed with these substances before they are introduced into the tube.	[1] Salts of tin for the most part, especially stannous chloride, are reduced without the aid of fluxes. This should be tried, as success denotes proficiency in the use of the blow-pipe. The globules, dissolved in hydrochloric acid and treated with mercuric chloride, give a white ppt. (See Group II. Sect. i. Tin, iv.)	[1] The metal volatilizes, oxidises, and condenses to form this incrustation. [2] The mass first fuses to a yellow liquid.	[1] The crystals are minute octahedra. [2] The current of air being excluded when the metal volatilizes, oxidation cannot take place, hence the results in this experiment (α) depend on the substance with which the metal is combined.	[1] The metal is reduced, and being infusible appears as a black mass.	

GROUP II. Section ii.

	BISMUTH. Subnitrate of Bismuth.	**CADMIUM.** Chloride of Cadmium.	**COPPER.** Sulphate of Copper.
Heated on Charcoal. i. In the *outer* flame.	**Fuse to a Brown mass.** On cooling becomes Yellow.		
ii. In the *inner* flame.	**Metallic globules**[1]. *brittle.* **With yellow incrustation.**	**Red-brown incrustation**[1].	**Metallic particles**[1]. *malleable.*
Heated on Platinum Wire with Borax. i. In the *outer* flame.	**Coloured bead**[2]. **Yellow** (hot). **Colourless** (cold).	**Transparent clear bead.** If *saturated* with Cadmium salt, becomes *milk-white* on cooling.	**Coloured bead.** **Green** (hot). **Blue** (cold).
ii. In the *inner* flame.			**Colourless** (hot). **Brick-red**[2] (cold).
	[1] The reduction of metal is facilitated by the use of carbonate of sodium. The incrustation is also more apparent when this flux is employed, since metallic bismuth is more volatile than the oxide. [2] Very small quantities of bismuthic salt should be used, as otherwise the bead becomes opaque. *Generally in all experiments with Borax on Platinum Wire, very small quantities of the salts to be tested should be employed, unless otherwise recommended.*	[1] The metal is reduced, volatilises, re-oxidises as it passes through the outer flame, and condenses on a cool part of the charcoal as red-brown oxide.	[1] The reduction of metal is facilitated by the use of carbonate of sodium. The colour (copper-colour) of the particles is characteristic. If washed in a mortar, the colour and malleability become very apparent. [2] It is necessary here that the bead should be fully saturated with the salt of copper.

GROUP III.

	ALUMINIUM. Sulphate of Aluminium.	**CHROMIUM.** Sesquichloride of Chromium.	**IRON.** Ferrous Salts Protosulphate of Iron. Ferric Salts Perchloride of Iron.
Heated on Charcoal. i. In the *outer* flame. α. Alone.	Infusible. Incandescent mass.	Infusible. Green residue[1].	Infusible. Coloured residu[e] Brown (hot). Red (cold).
β. Residue moistened with $CoNO_3$.	Blue coloration[1].		
ii. In the *inner* flame.			Infusible. Black residue
Heated on Platinum Wire with Borax. i. In the *outer* flame.	Colourless bead.	Green bead.	Coloured bead[1]. Red (hot). Yellow, or Colourless (cold).
ii. In the *inner* flame.		Green bead.	Slightly Yellow (hot). Bottle-green, or Colourless (cold).
	[1] Intense heat must be applied to produce this coloration, which is extremely characteristic.	[1] Both chromous and chromic salts behave in this manner. The chromous having this tendency to pass into a higher state of oxidation, are powerful reducing agents.	[1] The variations in colour depen[d] on the quantity of iron salt present.

GROUP IV.

	MANGANESE. Sulphate of Manganese.	ZINC. Sulphate of Zinc.	COBALT. Nitrate of Cobalt.	NICKEL. Sulphate of Nickel.
…ated on Charcoal. In the *outer* flame. a. Alone.	Infusible. Brown residue.	Infusible. Incandescent residue[1]. Yellow (hot). White (cold).	Infusible. Dark-green powder.	Infusible. Greenish o Dull-black residue.
Residue moistened with $CoNO_3$.	———	Green coloration.	———	———
In the *inner* flame.	———	White incrustation[2] (distant.)	———	———
…sed on Plati…m Wire with Borax. In the *outer* flame.	Coloured bead. Amethyst.	Clear bead[3].	Coloured bead[1]. Blue.	Coloured bead. Violet (hot). Red-brown (cold).
In the *inner* flame.	Colourless[1].	Milk-white[4].	Blue.	Grey[1].
…sed on Plati…m Foil with …aCO_3 + KNO_3.	Bluish-green coloration (cold).			
	[1] Care must be taken to use an extremely minute quantity of manganese salt. The bead assumes a pink colour before becoming colourless.	[1] The brilliant incandescence is characteristic. The coloration only appears on removal from the flame. [2] As has been pointed out in analogous cases, reduction of metal first takes place, the metal volatilising, oxidising, and condensing on a cool part of the charcoal support. [3] If saturated with the zinc salt. [4] Intense heat should be avoided, or reduction of metal takes place, and the metal alloys with the platinum.	[1] By the colour of this bead, cobalt may be recognized even in the presence of other metals.	[1] On adding nitrate of potassium to this bead, and heating again *in the oxidising flame*, it assumes a purple tint.

GROUP V.

	BARIUM. Nitrate of Barium.	**STRONTIUM.** Nitrate of Strontium.	**CALCIUM.** Chloride of Calcium.
Heated on Charcoal. In the *outer* flame. a. Alone.	Infusible residue[1]. Slightly incandescent.	Infusible residue[1]. Slightly incandescent.	Infusible residue[1]. Highly incandescent[2].
β. Residue moistened with $CoNO_3$.	—	—	—
Heated alone on Platinum Wire.	Flame is coloured[2]. Yellowish-green.	Flame is coloured[2]. Crimson.	Flame is coloured. Orange-red[3].
	[1] Chlorides, bromides, and iodides of barium resist decomposition into the oxide, and hence remain as a fused mass upon the charcoal. [2] An alcoholic solution of a barium salt manifests this reaction equally well.	[1] Chlorides, bromides, and iodides of strontium behave like the corresponding salts of barium. [2] An alcoholic solution burns with a crimson flame. The colour is perceptible in presence of a barium salt.	[1] Chlorides, bromides, and iodides remain undecomposed. [2] The incandescence is observed both in the case of the oxide or undecomposed chloride, bromide, and iodide. [3] Pure calcium gives a *red* flame, the orange tint being due to a trace of sodium. The flames of strontium and calcium should be carefully compared, as some confusion is liable to occur in determining between the colours of these flames. Calcium cannot be detected by the coloration of its flame, in presence of strontium.

GROUP VI.

POTASSIUM. Nitrate of Potassium.	SODIUM. Sulphate of Sodium.	AMMONIUM. Chloride of Ammonium.	LITHIUM. Chloride of Lithium.
Fuse into the charcoal. no!	Fuse into the charcoal.	Volatile. White incrustation.	Fuse into the ch coal.
Flame is coloured[1]. Violet.	Flame is coloured[1]. Yellow.		Flame is coloured Carmine[1].
1 Part of the salt volatilizes since certain salts of potassium, especially the chloride and nitrate, volatilize at an intense red heat; and is decomposed by the carbonaceous constituents of the flame. Potassium is set free in the form of vapour and burns with the characteristic violet flame. The coloration is not perceptible to the naked eye in presence of sodium. The interposition of a piece of dark blue glass between the eye and the flame renders it perfectly visible, since the blue glass intercepts the yellow rays of the sodium flame.	1 Sodium salts are more volatile at an intense red heat than the corresponding potassium salts, hence the coloration of the sodium flame is more marked.		1 The colour of this flame sembles that of strontium, differs from it in contain less yellow. Presence of salts of sodi obscures this reaction. Po sium does not materially in fere.

THE PROCESS OF SOLUTION.

(1) All bodies must be reduced to the liquid state (if not already in that condition) before the experiments in the following tables can be performed.

(2) The following solvents are usually employed in analysis, and are applied successively in the order here given:—Water. Hydrochloric Acid. Nitric Acid. Nitro-hydrochloric Acid.

(3) The substance must be reduced to a fine powder in a mortar, before proceeding to dissolve it, in order to facilitate the action of solvents.

(4) The solubility or otherwise of any solid in a solvent may be ascertained by boiling the solid in the liquid, allowing the undissolved residue (if any) to subside, decanting the supernatant liquid, and evaporating a portion of the latter on platinum foil. If a residue remains on the foil, the solid may be deemed sufficiently soluble for all ordinary purposes.

(5) Decantation should always be preferred to filtration, in the process of solution, if possible.

(6) Before treating any substance with acids, the student should endeavour to determine the particular acid which is likely to be applicable to any given salt, as an inference from the preliminary examination with the blowpipe, or by a reference to the tables of reaction.

e.g. If Silver, Lead, or Mercury, have been detected by the blowpipe experiments, and water fails to dissolve the salt operated upon, *Nitric Acid* must be employed as the solvent (Gr. I. col. i).

(7) The above acids should be employed in a *tolerably concentrated form* and in *small quantity*. Any required amount of dilution can readily be obtained after boiling with the strong acid. Some bodies dissolve in dilute acids which will not dissolve in strong acids, whilst other bodies dissolve in the latter and are reprecipitated on dilution. In the latter case the substance must be treated as directed in the following paragraph.

(8) If a body resists solution in water, hydrochloric acid, nitric acid, and nitro-hydrochloric acid, it must be well mixed with four or five times its weight of dry carbonate of potassium and sodium, fused in a porcelain crucible (over a gas burner or in a furnace) and allowed to cool. The fused mass must then be treated with water, and heated. If water fails to dissolve it, the acids mentioned in (2) must be employed, but in all cases the mass must be well washed with water in order to remove any soluble bodies which may be present.

(9) In the case of compounds of cyanogen, in order to guard against possible contingencies, the substance should be decomposed by boiling with solution of hydrate of potassium, and some carbonate of sodium; the residue washed and dissolved in acid.

BEHAVIOUR OF BASIC RADICALS WITH REAGENTS IN THE WET WAY.

GROUP I.

METALS WHICH ARE PRECIPITATED FROM SOLUTIONS OF THEIR SALTS, AS CHLORIDES, BY *HYDROCHLORIC ACID.*

	SILVER Nitrate of Silver.	**LEAD** Nitrate of Lead.	**MERCURY** (Mercurous salts). Sub-nitrate of Mercury.	
I. **Hydrochloric Acid.**	$AgNO_3 + HCl$ = white (curdy) α .... $AgCl$ insol. in water. sol. in hydrate of ammonium, reppt. by nitric acid.	$PbNO_3 + HCl$ = white .... $PbCl$ sol. in boiling water α. sol. slightly in hydrate of ammonium.	$Hg_2NO_3 + HCl$ = white .... Hg_2Cl = Calomel insol. in water. insol. in hydrate of ammonium, but blackened α.	I. Chlorides.
II. **Hydrosulphuric Acid.**	$2AgNO_3 + H_2S$ = black β .... Ag_2S insol. in alkaline sulphides and hydrate of ammonium. sol. in concd. nitric acid.	$2PbNO_3 + H_2S$ = black .... Pb_2S insol. in alkaline sulphides and hydrates. sol. in boiling nitric acid β.	$2Hg_2NO_3 + H_2S$ = black β .... $(Hg_2)_2S$ insol. in sulphide of ammonium. sol. in nitro-hydrochloric acid.	II. Sulphides.
III. **Hydrate of Ammonium.**	$AgNO_3 + NH_4HO$ = olive-green γ .... $AgHO$ *add HCl as soon as the ppt. is formed* δ. sol. in excess of hydrate of ammonium.	$4PbNO_3 + 4NH_4HO$ = white γ .... $2(PbHO).Pb_2O$ insol. in excess of hydrate of ammonium. sol. in excess of hydrate of potassium.	$3Hg_2NO_3 + 3NH_4HO$ = black compound...... $NH.(Hg_2)_2NO_3$	III. Hydrates, &c.
IV. **Hydrate of Potassium.**	$2AgNO_3 + 2KHO$ = brown .... Ag_2O insol. in excess of hydrate of potassium.	$PbNO_3 + KHO$ = white .... $PbHO$ sol. in excess of hydrate of potassium, esp. on heating.	$2Hg_2NO_3 + 2KHO$ = brown-black .... $(Hg_2)_2O$ decomposes on boiling with excess of hydrate of potassium into mercuric oxide (Hg_2O) and metallic mercury.	IV. Oxides, &c.
V. **Chromate of Potassium.**	$AgNO_3 + KCrO_3$ = crimson ....$AgCrO_3$ sol. in hydrate of potassium, and nitric acid. insol. in water.	$PbNO_3 + KCrO_3$ = yellow .... $PbCrO_3$ sol. in hydrate of potassium. insol. in water and nitric acid.	$Hg_2NO_3 + KCrO_3$ = red γ .... Hg_2CrO_3 sol. slightly in water and nitric acid.	V. Chromates.

VI. Iodide of Potassium.	$AgNO_3 + KI$ = yellow (curdy)..............AgI sol. partially in excess of iodide of potassium.	$PbNO_3 + KI$ = yellow δ....................PbI sol. in excess of iodide of potassium.	$Hg_2NO_3 + KI$ = green-yellow (pulverulent) δ...Hg_2I sol. in excess of iodide of potassium.	VI. Iodides.
VII. Sulphuric Acid.	$2AgNO_3 + H_2SO_4$ = white (crystalline).........Ag_2SO_4 sol. in much water, and in concd. sulphuric acid, reppt. partially by water.	$2PbNO_3 + H_2SO_4$ = white (crystalline)......Pb_2SO_4 insol. in water. sol. in boiling hydrochloric acid, concd. sulphuric acid, and acetate of ammonium.	$2Hg_2NO_3 + H_2SO_4$ = white (crystalline)......$(Hg_2)_2SO_4$ sol. slightly in water. sol. in hot concd. sulphuric acid, in nitric acid, and reppt. by sulphuric acid.	**VII. Sulphates.**
VIII. Metallic Copper.	$AgNO_3 + Cu$ = $Ag + CuNO_3$ minute spangles ε, sometimes brown...Ag	———	$Hg_2NO_3 + Cu$ = $Hg_2 + CuNO_3$ grey deposit ε on surface of copper...Hg_2	VIII. Reduction of Metal.
	α On exposure to light, the ppt. becomes violet and finally black, due to the formation of a dark oxychloride, with simultaneous evolution of gaseous hydrochloric acid. In very dilute solutions, a bluish-white opalescence becomes immediately apparent on the addition of hydrochloric acid. β This sulphide constitutes the chief ore of silver technically known as "Silver glance." γ The colour changes rapidly to brown, since the hydrate rapidly decomposes into the brown oxide and water. $2AgHO = Ag_2O + H_2O$. In an acid solution of a silver salt, hydrate of ammonium produces no ppt. δ Hydrochloric acid is added, because a compound termed "fulminating silver" is often deposited from the solution, and the argentic oxide is by this means converted into chloride. ε Many other metals also reduce silver from solutions of its salts—e.g. mercury, zinc.	α After dissolving in boiling water, the chloride again separates from the solution on cooling, in the form of acicular crystals*. β By concentrated nitric acid, this ppt. is converted into the white Plumbic sulphate, with separation of sulphur. Sulphide of lead occurs native as "galena." γ The ppt. is more usually a basic salt, containing some of the acid radical of the original plumbic salt. By heat it is converted into plumbic oxide (Pb_2O). δ This ppt. is soluble in much boiling water, separating from the solution on cooling in the form of golden yellow scales†. * In actual analysis, if the solution is dilute, lead when present is only partially or not at all precipitated. In any case sufficient lead remains in solution to be thrown down as sulphide by hydrosulphuric acid, and hence will be detected on application of this acid as a group-test. † If the least trace of bismuth is present the scales are orange or crimson.	α The white chloride suffers decomposition, similar to that which the nitrate experiences (see III)—the black compound containing chlorine instead of the nitric radical. $3Hg_2Cl + 3NH_4HO = NH_2(Hg_2)_2Cl + 2NH_4Cl + 3H_2O$. β Boiled with water, this sulphide breaks up into mercuric sulphide (Hg_2S) and metallic mercury. γ The ppt. becomes black on exposure to light. Often instead of the neutral salt, whose formula is given above, basic salt is precipitated, having the formula ($3Hg_2CrO_4.(Hg_2)_2O$). Nitric acid converts the latter into the neutral salt. δ This ppt. usually contains, besides, some mercuric iodide (HgI). ε Amongst other metals which reduce mercury from solutions of mercurous salts, are zinc—tin as chloride—iron, as ferrous sulphate.	

GROUP

METALS WHICH ARE PRECIPITATED FROM SOLUTIONS

THE PRECIPITATE *SOLUBLE*

	TIN (Stannous salts). Proto-Chloride of Tin.	**TIN** (Stannic salts). Bichloride of Tin.	**ANTIMONY** (Antimonious sa Terchloride of Antimony.
I. **Hydrosulphuric Acid.**	$SnCl + H_2S$ = dark-brown .... Sn_2S sol. in sulphide of ammonium α, reppt. by hydrochloric acid β. sol. in hydrate of potassium, reppt. unchanged by hydrochloric acid. insol. in carbonate of ammonium.	$2SnCl_2 + 2H_2S$ = yellow α .... Sn_2S_2 sol. in sulphide of ammonium, reppt. by hydrochloric acid. sol. in hydrate of potassium. insol. in carbonate of ammonium.	$2SbCl_3 + 3H_2S$ = orange (characteristic) α...Sb_2S_3 sol. in sulphide of ammonium. sol. in hydrate of potassium. insol. in carbonate of ammonium.
II. **Hydrate of Potassium.**	$SnCl + KHO$ = white .... $SnHO$ sol. in excess of hydrate of potassium. On boiling this solution, metallic tin is separated γ.	$2SnCl_2 + 4KHO$ = white .... $H_2Sn_2O_3$ stannic acid β. sol. in excess of hydrate of potassium.	$2SbCl_3 + 6KHO$ = white (flocculent) .... Sb_2O_3 sol. in excess of hydrate of potassiu insol. in hydrate of ammonium.
III. **Water.**	——	$2SnCl_2 + 3H_2O$ = white .... $H_2Sn_2O_3$ stannic acid.	$12SbCl_3 + 15H_2O$ = white .... $2SbCl_3 . 5S$ powder of algaro sol. in tartaric acid γ.
IV. **Mercuric Chloride.**	$SnCl + 2HgCl$ = white .... Hg_2Cl decomposes when heated with excess of stannous chloride, with separation of metallic mercury δ.	——	——
V. **Cupric Sulphate.**	——	——	——
VI. **Nitrate of Silver.**	——	——	——

α The ppt. is only partially soluble in ordinary sulphide of ammonium [$(NH_4)_2S$]—it only dissolves completely when a higher sulphide is used $(NH_4)_2S_2$.
β Hydrochloric acid reprecipitates as stannic sulphide (Sn_2S_2).
γ Part of the tin becomes oxidised at the expense of the other part—thus:
$4SnHO + 2KHO = K_2Sn_2O_3 + 2H_2O + Sn_2$.
δ Stannous chloride removes all the chlorine from the mercury, and metallic mercury separates as a grey deposit. Hence the value of stannous chloride as a reducing agent.

α Alkaline solutions are not precipitated. If the stannous chloride is in excess the ppt. is white.
β The salts of stannic acid have the general formula $M_2Sn_2O_3$.
Another modification of the hydrate exists, as *Metastannic acid*, having the formula $H_{10}Sn_{10}O_{15}$. It is produced by the action of concentrated nitric acid upon metallic Tin, as a white powder. Its salts have the general formula $M_2H_8Sn_{10}O_{15}$.

α This ppt. only forms completely i acid solution.
β Sb_2O_3 should be dissolved in hydroch acid, evaporated to a small bulk, and poured into much water, when the white chloride will be precipitated*.
γ This reaction distinguishes antimon oxychloride from bismuthic oxychloride Bismuth, Gr. II. Sect. ii.).

* This ppt. will often appear on simpl luting solutions of antimonious salts may then be removed by hydrochloric aci

EIR SALTS, AS SULPHIDES, BY *HYDROSULPHURIC ACID.*

ʟPHIDE OF AMMONIUM.

ɪTIMONY (Antimonic salts). Antimoniate of Potassium.	**ARSENIC** (Arsenious salts). Arsenious Acid.	**ARSENIC** (Arsenic salts). Arsenic Acid.	
$_2SbO_3$ α + $5H_2S$ = **ınge-yellow** β............Sb_2S_5 in sulphide of ammonium, reppt. by hydrochloric acid. in hydrate of potassium γ.	$2H_3AsO_3$ α + $3H_2S$ = **orange** β....................As_2S_3 sol. in sulphide of ammonium, reppt. by hydrochloric acid. sol. in hydrate of potassium. sol. in carbonate of ammonium.	$2H_3AsO_4 + 5H_2S$ = **yellow** α......................As_2S_5 sol. in sulphide of ammonium. sol. in hydrate of potassium. sol. in carbonate of ammonium.	I. Sulphides.
——	—— γ	—— β	II. Hydrates and Oxide.
$l_5 + 4H_2O$ = **hite**.......................H_3SbO_4 metantimonic acid δ. in hydrate of ammonium. in water, reppt. by acids.	——	——	III. Hydrate and Oxychloride.
——	——	——	IV. Mercurous Chloride.
——	$2H_3AsO_3$ + $6NH_4HO$ } + $3Cu_2SO_4$ δ = **green**........................$2Cu_3AsO_3$ ε sol. in hydrate of ammonium and acids.	$2H_3AsO_4$ + $6NH_4HO$ } + $3Cu_2SO_4$ = **bluish-green**.............$2Cu_3AsO_4$ sol. in hydrate of ammonium.	V. Cupric Arsenites and Arseniates.
——	H_3AsO_3 + $3NH_4HO$ } + $3AgNO_3$ = **yellow**.......................Ag_3AsO_3 sol. in hydrate of ammonium, nitric acid, and acetic acid.	$H_3AsO_4 + 3AgNO_3$ = **brick-red** γ.................Ag_3AsO_4 sol. in hydrate of ammonium, nitric acid and acetic acid.	VI. Argentic Arsenites and Arseniates.

f the acid metantimoniate of potassium, ranular antimoniate (such as is employed sting for sodium) is used, the formula $_2Sb_2O_7$ must be substituted for the above. Presence of tartaric acid prevents pre-ation.

ntimoniate of potassium separates on ling, as a crystalline ppt.

trictly the above reaction consists of two s, an oxychloride of antimony being first ed thus:—$SbCl_5 + H_2O = Cl_3SbO + 2HCl$

and

$Cl_3SbO + 3H_2O = H_3SbO_4 + 3HCl.$

ntimonic acid is very unstable, rapidly ing into antimonic acid.

$H_3SbO_4 - H_2O = HSbO_3.$

ce in many works the product of the re-n of penta-chloride of antimony with r, is called antimonic acid.

ere are three modifications of antimonic ate—

ntimonic Acid ($HSbO_3$)—formed from llic antimony and aqua regia.

etantimonic Acid (H_3SbO_4) formed as e.

rantimonic Acid ($H_4Sb_2O_7$) formed by de-posing an alkaline antimoniate by an acid. is the metantimonic acid of Fremy and r chemists, forms two classes of salts b_2O_7 and $M_2H_2Sb_2O_7$. It is analogous s properties to metantimonic acid.

is series is analogous to the phosphoric s. (See Odling's *Manual*, p. 356.)

α Arsenious acid is only known in the state of aqueous solutions. It may be prepared by boiling arsenious anhydride (white arsenic) in water, on cooling a portion of the latter is held in solution.

β Alkaline solutions are not precipitated: the above ppt. only occurs in presence of a free acid, such as hydrochloric acid.

γ The oxide is obtained when ores of metals containing arsenic are roasted. It occurs in two varieties, the vitreous and crystalline.

δ The aqueous solution of arsenious acid used in this and the following experiment, must be exactly neutralised by hydrate of ammonium, otherwise no effect is produced.

ε Known as Scheele's green.

α The ppt. does not form immediately, even in a concentrated acidified solution: prolonged passage of hydrosulphuric acid is required to throw down the sulphide. The ppt. is most completely formed by boiling arsenic acid solution with sulphurous acid until no odour is evolved, thus producing arsenious acid ($H_3AsO_4 + H_2SO_3 = H_3AsO_3 + H_2SO_4$), then passing hydrosulphuric acid gas.

The sulphide as above seems to be merely a mixture of one atom of tersulphide (As_2S_3) and two atoms of sulphur.

β The oxide is obtained by oxidising arsenious anhydride (As_2O_3) or arsenious acid (H_3AsO_3) by means of nitric acid. It occurs in the form of long prisms, containing $2H_3AsO_4 + H_2O$.

γ The ppt. is more completely formed when the solution of arsenic acid is neutralised by hydrate of ammonium.

GROUP II. SECTION I. (Continued.)

METALS WHICH ARE PRECIPITATED FROM SOLUTIONS OF THEIR SALTS, AS SULPHIDES, BY *HYDROSULPHURIC ACID.*

THE PRECIPITATE *SOLUBLE* IN SULPHIDE OF AMMONIUM.

	PLATINUM. Bichloride of Platinum.	**GOLD.** Terchloride of Gold.	
I. **Hydrosulphuric Acid.**	$2PtCl_2 + 2H_2S$ = **brown-black** α............Pt_2S_2 sol. in sulphide of ammonium, reppt. by hydrochloric acid.	$2AuCl_3 + 3H_2S$ = **black**...............Au_2S_3 sol. in sulphide of ammonium, reppt by hydrochloric acid. sol. in hydrate of potassium.	I. Sulphides.
II. **Hydrate of Potassium.**	$PtCl_2(+HCl) + KHO$ β = **yellow** (see IV)............$KPtCl_3$	The formula of the hydrate, produced by the action of hydrate of potassium not in excess on auric chloride, is not known α.	II. Chloro-Platinate.
III. **Iodide of Potassium.**	$PtCl_2 + 2KI$ = **brown** γ............PtI_2 sol. in alcohol, forming a yellowish-green solution.	$AuCl_3 + 3KI$ = **green** β..............AuI_3	III. Iodides.
IV. **Chloride of Potassium.**	$PtCl_2 + KCl$ = **yellow** δ............$KPtCl_3$ insol. in alcohol.	——	IV. Chloro-Platinate.
V. **Ferrous Sulphate.**	——	$AuCl_3 + 3Fe_2SO_4$ γ = $(Fe_2)_2(SO_4)_3 + Fe_2Cl_3$ **brown powder** δ............Au	V. Reduction of Metal.
VI. **Stannous Chloride.**	The addition of stannous chloride to a solution of platinic chloride, containing excess of hydrochloric acid, gives a brown solution ε.	$AuCl_3(+HCl) + SnCl$ = **purple**............$AuSn_3O_5$? Purple of Cassius ε.	VI. Purple of Cassius.

α The ppt. does not usually form immediately, heat promotes its formation. In an alkaline solution the precipitation is only partial.

β The hydrate is only produced when hydrates of the alkalies are added to a solution of platinic nitrate.

γ The ppt. only forms upon standing or heating. The solution is coloured immediately.

δ In dilute solutions the liquid, after adding chloride of potassium, must be evaporated to dryness, and the residue treated with alcohol.

ε The coloration is due to the reduction of platinic chloride to the condition of platinous chloride (PtCl).

α If hydrate of potassium in excess is added to a solution of auric chloride no ppt. is produced, but on adding tannic acid to the clear solution, a black ppt. is formed (Au_2O).

β The ppt. dissolves on agitation.

γ Excess of hydrochloric acid must be present, otherwise an insoluble salt of iron would be formed.

δ By means of ferrous sulphate, gold may be detected in extremely dilute solutions by the violet or blue tint which is imparted to them on addition of this reagent.

If the above ppt. is dried, it exhibits metallic lustre when rubbed.

ε When the quantity of gold present is very minute, a dusky red tinge pervades the solution. The reaction is more delicate if stannous chloride contains some stannic chloride. The best way of applying the test is as follows. The solution to be tested should be placed with much water in a beaker and acidified with a few drops of nitric acid or ferric chloride. The stannous chloride, containing a trace of stannic chloride and acidified with hydrochloric acid until quite clear, should then be poured slowly into the solution to be tested, when the characteristic colour will mark the course of the precipitant through the solution.

In actual analysis, a special and separate examination of a portion of the original solution is made, when the presence of one or both of the above metals is suspected.

SPECIAL TESTS FOR ARSENIC.

I. MARSH'S TEST.

BEHAVIOUR OF ANTIMONY AND ARSENIC BY THIS METHOD, CONTRASTED.

Experiment.	Result. Antimony.	Result. Arsenic.
Hydrogen gas is generated from zinc and sulphuric acid (both previously proved to be perfectly free from arsenic). The solution to be tested is introduced into the generating apparatus.	Sb unites with *nascent* hydrogen forming *Antimoniuretted Hydrogen.* $SbCl_3 + 6H = 3HCl + SbH_3$.	As unites with *nascent* hydrogen, forming *Arseniuretted Hydrogen.* $As_2O_3 + 12H = 3H_2O + 2AsH_3$.
The gas (SbH_3 or AsH_3) is dried.	By passing through *sulphuric acid,* or over *chloride of calcium.*	By passing over *chloride of calcium.* (Sulphuric acid decomposes AsH_3.)
A fine jet of hard glass is attached to the apparatus, at the end of the drying tube, and the escaping gas (SbH_3 or AsH_3) is ignited (after allowing all the air to be displaced).	Burns with a *bluish-green flame; white fumes* arise $= Sb_2O_3$.	Burns with a *bluish-white flame; white fumes* arise $= As_2O_3$; *garlic odour.*
A cold porcelain surface is depressed into the flame.	*A black spot* of metallic antimony appears, which is lustrous if thin α.	*A black lustrous spot* of metallic arsenic appears α.
The metallic spot is treated with nitric acid, which is carefully evaporated, and then nitrate of silver is added.	No change of colour occurs.	Colour changes to yellow or red. If oxidation has given rise to *arsenious acid,* it will change to *yellow;* if to *arsenic acid,* it will change to *red.* (See Table for Group II. Sect. 1.)
The metallic spot is treated with solution of hypochlorite of sodium α.	The spot scarcely at all dissolves β.	The spot immediately dissolves.
The spot is moistened with solution of sulphide of ammonium.	The spot immediately dissolves.	The spot remains undissolved.
That portion of the jet-tube nearest the $CaCl_2$ is heated by the blowpipe flame.	*A lustrous mirror* appears on the inside of the tube beyond where the blowpipe flame impinges γ.	*A lustrous mirror* appears on the inside of the tube, *at some distance beyond* where the blowpipe flame impinges β.
The tube containing the mirror is detached from the hydrogen apparatus and attached to one in which pure H_2S is being generated (washed and dried by passing through H_2SO_4); gentle heat is at the same time applied to the mirror.	The mirror becomes of a *reddish-yellow* or *brownish-black* colour δ.	The mirror becomes *lemon-yellow* γ.
The tube with the altered mirror is attached to an apparatus from which HCl is evolved (by heating a concentrated solution and drying by means of H_2SO_4).	The Sb_2S_3 entirely disappears ε.	No change occurs.
The volatile $SbCl_3$ is conducted into water, which dissolves it. H_2S is passed through this solution.	*Orange ppt.* of Sb_2S_3.	

α Really $NaClO + NaCl$, prepared by mixing a solution of CaCl and Na_2CO_3 in excess, and filtering.

α The cold surface reduces the temperature, hence a portion of the antimony or arsenic is not oxidised (as it is when burning, giving rise to white fumes), but condenses on the cold surface.
β Antimony cannot be detected by this means, in the presence of arsenic.
γ Heat decomposes SbH_3 into its constituents, and metallic Sb is deposited, the gas escaping at the jet being almost pure H.
δ This is due to the conversion of the metallic Sb into Sb_2S_3.
ε It is converted into volatile, colourless $SbCl_3$.

α If the porcelain be allowed to remain in the flame more than a second or two, the spot disappears, from volatilisation of the metal.
β As is more volatile than Sb, hence is deposited in a cooler part of the tube.
γ Due to formation of As_2S_3. (See Table for Group II. Sect. 1.)

This test will only serve to distinguish whether one or both of these metals are present in a substance. It will not indicate one, with certainty, in the presence of the other. If arsenic *alone* is present it is decisive.

II. REINSCH'S TEST.

Clean metallic copper when boiled with a solution, containing arsenic or its com**pounds, acidified with hydrochloric acid, reduces** the arsenic to the metallic state. The latter forms a *steel-grey film* upon the copper, or if the quantity of arsenic is considerable and the boiling is prolonged, it appears as *large black scales.* The film or scales may be sublimed and so converted into arsenious **acid, which appears as** iridescent octahedra. **This** may be further tested **by boiling in water and treating** with hydrosulphuric **and** hydrochloric **acid gas, or by nitrate of silver.**

METHOD OF APPLICATION OF REINSCH'S TEST.

Experiment.	Result.
The solution to be tested is boiled with ⅛th its bulk of hydrochloric acid.	
Pieces of copper wire (an inch long) previously cleaned with concentrated nitric acid and washed, are boiled with the hydrochloric acid solution, for two or three minutes.	Arsenic, if present, is reduced on the surface of the copper, either appearing as a *grey discoloration* or film, or as large *black scales.*
The copper is removed, washed with distilled water, dried between bibulous paper, or in a water-bath, introduced into a hard glass tube contracted at one end, and heated by holding it obliquely in a gas flame.	A *crystalline sublimate* of arsenious acid forms on the cool part of the tube.
The portion of the tube containing the sublimate is filed off and boiled in water.	Solution of arsenious acid is formed.
A portion of the solution is tested by hydrosulphuric acid gas.	*Yellow tersulphide of arsenic* is precipitated.
Through the same solution hydrochloric acid gas is passed.	No change takes place.
A second portion is tested with nitrate of silver, (taking care to neutralize first with ammonia).	*Yellow arsenite of silver* is precipitated.

III. FRESENIUS' AND BABO'S TEST.

If arsenites, arsenious acid, or tersulphide of arsenic (the latter is preferable when merely trying the reaction) are fused with equal parts of dry carbonate of sodium and cyanide of potassium, the whole of the arsenic is reduced to the metallic state. Being extremely volatile, it condenses again, if the operation is conducted in a bulb-tube, upon a cool part of the tube, forming the characteristic arsenical mirror. If the operation is conducted in an apparatus which ensures the preservation of the arsenical vapours from contact with the atmosphere, the delicacy of the test is so far increased as to render it at once the least objectionable in its application and the most efficacious as a means of detecting the presence of arsenic.

Fresenius and Babo instituted a series of experiments which showed that the required delicacy could be obtained by heating the mixture of the arsenical compound, carbonate of sodium and cyanide of potassium, in a stream of carbonic acid gas (anhydride).

METHOD OF APPLICATION OF FRESENIUS' AND BABO'S TEST.

Experiment.	Result.
Carbonic acid gas is generated in an ordinary gas apparatus, from *lumps* of limestone or marble and hydrochloric acid.	
The gas is dried by passing it through sulphuric acid.	
The mixture (3 parts dry carbonate of sodium, 1 part dry cyanide of potassium, 1 part dry tersulphide arsenic, intimately mixed in a mortar) is introduced into a combustion-tube of hard glass, drawn out at one end and attached at the large end to the egress tube of the wash-bottle.	
The gas is allowed to pass so as to expel all air, and then at the rate of one bubble per second (this can be effected by pouring water into the generating bottle).	
The combustion tube is heated in its entire length, to expel all moisture, then to redness at the shoulder beyond the mixture, and then at the same time the mixture until it is entirely fused.	The vapours of metallic arsenic are carried onwards by the carbonic acid gas, and the shoulder of the tube being heated, they pass on and condense in the drawn-out portion as a *lustrous mirror* [a].
	A garlic odour may be detected at the end of the tube.
The portion of the tube containing the mirror may be cut off with a file and after dissolving in water, may be further tested by means of nitrate of silver, as in the case of Reinsch's or Marsh's test.	*No compound of antimony whatever yields a similar reaction under the circumstances.*

[a] If only a minute quantity of Arsenic is present, the mirror appears as a *thin grey film*.

GROUP II.

METALS WHICH ARE PRECIPITATED FROM SOLUTIONS OF
THE PRECIPITATE *INSOLUBLE*

	MERCURY (Mercuric salts). Perchloride of Mercury.	**BISMUTH.** Terchloride of Bismuth.
I. ***Hydrosulphuric Acid.**	$6HgCl + 2H_2S$ = **white**α ....$2(HgCl.Hg_2S)$ insol. in sulphide of ammonium. insol. in nitric acid.	$2BiCl_3$α + $3H_2S$ = **brown-black**............Bi_2S_3 insol. in sulphide of ammonium. sol. in nitric acid.
II. **Hydrates of the Alkalies.**	$HgCl + KHO$β = **reddish-brown**γ......$HgHO$ (if only a small quantity of KHO is added.) insol. in excess of hydrate of potassium; *on adding the excess, the ppt. is converted into the yellow oxide* (Hg_2O). insol. in hydrate of ammonium.	$3BiCl_3 + 6KHO$ = **white** (flocculent).........$BiH_2O_2.Bi_2O_3$ insol. in excess of hydrate of potassium β. insol. in hydrate of ammonium.
III. **Hydrate of Potassium.**	$4HgCl + 2KHO$ = **yellow**δ...............$2Hg_2O$ insol. in excess of hydrate of potassium.	Action of Water on Bismuthic Salts. $3BiCl_3 + 3H_2O$ = **white**γ... .... $BiCl_3.Bi_2O_3$ insol. in tartaric acid (see Antimony, p. 24).
IV. **Iodide of Potassium.**	$HgCl + KI$ = **scarlet**..............HgI sol. in excess of iodide of potassium.	$BiCl_3 + 3KI$ = **brown** (cystalline)......BiI_3 insol. in excess of iodide of potassium.
V. **Chromate of Potassium.**	$3HgCl + H_2O + KCrO_4$ = **yellow**...........$Hg_2O.HgCrO_4$ sol. in acids.	$BiCl_3 + 3KCrO_4$ = **lemon-yellow**........$Bi(CrO_4)_3$ sol. in nitric acid and hydrate of ammonium. insol. in hydrate of potassium δ.
VI. **Stannous Chloride. Metallic Iron.**	$2HgCl + SnCl$ = $SnCl_2 + Hg_2Cl$ **white ppt.**.........Hg_2Cl $Hg_2Cl + SnCl$ = $SnCl_2 + Hg_2$ **grey deposit**ε.........Hg	——
VII. **Cyanide of Potassium.**	Soluble ζ.	$BiCl_3 + 3KCy$ = **white**..............$BiCy_3$ insol. in excess of cyanide of potassium.

α The white ppt. is only produced by *small quantities* of the reagent. If the experiment is carefully performed, the colour will become, according to the amount of reagent added, yellow, orange, brown, and finally black from formation of the black sulphide (Hg_2S). The white ppt. has the formula assigned to it.

β The action of hydrate of ammonium on mercuric salts is to form a *double chloride and amide of mercury* ($HgCl.HgNH_2$) thus;

$2HgCl + 2NH_4HO = (HgCl.HgNH_2) + NH_4Cl + 2H_2O$.

The ppt. is known by the name of "white precipitate."

γ The hydrate rapidly passes into the oxide (Hg_2O). The experiment as detailed in II. must be carefully performed or the reactions will not occur.

δ When mercuric oxide is prepared in any other way than by the action of an alkali, it is red.

ε This reduction of metal takes place when mercuric chloride is boiled with excess of stannous chloride. The deposit if boiled with hydrochloric acid takes the form of globules.

ζ With mercuric nitrate, cyanide of potassium gives a white ppt. soluble in excess.

α A solution of bismuthous nitrate in water and hydrochloric acid may be substituted for bismuthic chloride.

β If after adding excess of hydrate of potassium, the ppt. is allowed to subside, the supernatant liquid decanted, and the residue evaporated to dryness, the ppt. will be converted into *Yellow Oxide* (Bi_2O_3).

The yellow oxide is not formed on merely boiling the ppt.

γ These salts are also termed *basic* salts. The chloride is usually employed in this reaction, as its decomposition is most complete; with nitrate, a mixed hydrate and nitrate of bismuth results.

$3Bi(NO_3)_3 + 6H_2O = [Bi(NO_3)_3.2BiH_3O_3] + 6HNO_3$.

δ The solubility of bismuthic chromate in nitric acid and its insolubility in hydrate of potassium, distinguish it from plumbic chromate.

SECTION II.

THEIR SALTS AS SULPHIDES BY *HYDROSULPHURIC ACID.*
IN SULPHIDE OF AMMONIUM.

CADMIUM. Chloride of Cadmium.	**COPPER.** Sulphate of Copper.	
$2CdCl + H_2S$ = **brilliant yellow**........Cd_2S insol. in sulphide of ammonium. sol. in nitric acid. insol. in cyanide of potassium α.	$Cu_2SO_4 + H_2S$ = **black** (in flakes) α..........Cu_2S insol. in sulphide of ammonium. sol. in nitric acid. sol. in cyanide of potassium.	**I.** **Sulphides.**
$CdCl + KHO$ β = **white**..........................CdHO insol. in excess of hydrate of potassium γ. sol. in hydrate of ammonium.	cold $Cu_2SO_4 + 2KHO$ β = **blue** (flocculent) γ............2CuHO insol. in excess of hydrate of potassium. *boiling converts this ppt. into the black oxide* (Cu_2O). sol. in hydrate of ammonium δ.	**II.** **Hydrates.**
——	boiling $Cu_2SO_4 + 2KHO$ = **black** ε..........................Cu_2O sol. in hydrochloric, nitric, or sulphuric acid, forming corresponding cupric salts.	III. Oxides.
soluble.	$Cu_2SO_4 + 2KI$ = I + **brown**........................Cu_2I sol. in excess of iodide of potassium.	IV. Iodides.
$CdCl + KCrO_4$ = **yellow**........................$CdCrO_4$	$Cu_2SO_4 + 2KCrO_4$ = **yellow-brown**..............$2CuCrO_4$ sol. in nitric acid and hydrate of ammonium, with the latter forming a green solution.	V. Chromates.
——	$Cu_2SO_4 + Fe_2$ = $Cu_2 + Fe_2SO_4$ **metallic coating on the iron**........Cu presence of an acid (HCl) accelerates this action.	VI. Reduction of Metal.
$CdCl + KCy$ = **white**......................CdCy sol. in excess of cyanide of potassium.	$Cu_2SO_4 + 2KCy$ = **green-yellow**................2CuCy sol. in excess of cyanide of potassium ζ.	VII. Cyanides.

α This reaction serves to distinguish between cadmium and copper in the process of analysis.
β Hydrate of ammonium also produces a white ppt. of hydrate with solutions of cadmium salts, but it is readily soluble in excess of the precipitant.
γ On treating this solution, as directed in note β, Bismuth, the *Brown Oxide* (Cd_2O) is formed.

α This ppt. rapidly absorbs oxygen from the air and passes into cupric sulphate:

$$Cu_2S+O_4=Cu_2SO_4$$

β Cupric sulphate produces with hydrate of ammonium a green-blue ppt. of basic sulphate of cuprammonium $[(NH_3Cu)_2SO_4]$. See note δ.
γ If the precipitant is deficient in quantity a green basic salt is formed.
δ Cupric hydrate dissolves in hydrate of ammonium, forming a blue solution. The metal is supposed to replace 1 eq. of hydrogen in the molecule of ammonium (NH_4), and so to constitute the basic radical termed "Cuprammonium" (NH_3Cu).
ε At a red heat cupric oxide gives up its oxygen to hydrogen or carbon—hence its value in organic analysis.
ζ Hydrochloric acid precipitates white cuprous cyanide (Cu_2Cy) from this solution, soluble in excess.

GROUP

METALS WHICH ARE PRECIPITATED FROM SOLUTIONS OF

	ALUMINIUM. Chloride of Aluminium.	CHROMIUM. Sulphate of Chromium or Chromic Acid.
I. **Hydrate of Ammonium.**	$Al_2Cl_3 + 3NH_4HO$ = **white** (gelatinous) α ...... $Al_2H_3O_3$ insol. in excess of hydrate of ammonium. sol. in hydrate of potassium β. insol. in chloride of ammonium.	$(Cr_2)_2(SO_4)_3 + 6NH_4HO$ = **bluish-green** .... $2Cr_2H_3O_3$ insol. in excess of hydrate of ammonium. sol. in hydrate of potassium. insol. in chloride of ammonium.
II. **Hydrate of Potassium.**	$Al_2Cl_3 + 3KHO$ = **white** .... $Al_2H_3O_3$ sol. in excess of hydrate of potassium, and reppt. by chloride of ammonium γ. *not reppt. on boiling.*	$(Cr_2)_2(SO_4)_3 + 6KHO$ = **green** .... $2Cr_2H_3O_3$ sol. in excess of hydrate of potassium, and reppt. by chloride of ammonium. *also reppt. on boiling.*
III. **Sulphide of Ammonium.**	$2Al_2Cl_3 + 3(NH_4)_2S = 6NH_4Cl + (Al_2)_2S_3$ and $(Al_2)_2S_3 + 6H_2O$ = **white ppt.** δ .... $2Al_2H_3O_3 + 3H_2S$.	A similar reaction occurs with chromic salts and sulphide of ammonium as is the case with salts of aluminium.
IV. **Ferri-Cyanide of Potassium.**	————	————
V. **Ferro-Cyanide of Potassium.**	The greenish ppt. usually produced by ferro-cyanide of potassium with solutions of aluminium salts, is a *hydrate* containing some cyanide of iron ε.	————
VI. **Sulphate of Potassium.**	$Al_2Cl_3 + 2K_2SO_4$ ζ = **white** .... $Al_2K(SO_4)_2$ = common alum η	soluble a.
VII. **Silicate of Potassium.**	$Al_2Cl_3 + 3KSiO_2$ = **white** .... $Al_2(SiO_2)_3$ ι.	————

α The ppt. dries into a transparent horny mass.

β When the ppt. is dissolved in hydrate of potassium, a new compound (aluminate of potassium) is formed, the aluminium forming part of the acid-radical.

$$Al_2H_3O_3 + KHO = KAl_2O_2.H_2O + H_2O.$$

The hydrate may be again obtained from the solution by adding chloride of ammonium, thus

$$KAl_2H_2O_3 + NH_4Cl = Al_2H_3O_3 + KCl + NH_3.$$

γ This reaction is explained in note β.

δ A sulphide is first formed in the reaction with sulphide of ammonium, which is immediately decomposed by water into the white hydrate and hydrosulphuric acid gas.

ε It will be observed that the hydrate is a constant result of reactions with even the most varied reagents and aluminium salts. The hydrate in this group generally is an insoluble salt, and its formation is very general.

ζ The solution of aluminic chloride and that of sulphate of potassium should be hot and concentrated (by evaporation).

η This salt is the type of the true alums. All true *alums* have 12 eqs of water of crystallisation:

$$[Al_2K(SO_4)_2 + 12aq].$$

a The soluble salt formed with sulphate of potassium is one of the alums (chrome-alum) having the formula $KCr_2(SO_4)_2 + 12aq$.

III.

THEIR SALTS, AS HYDRATES BY *HYDRATE OF AMMONIUM.*

IRON (Ferric salts). Perchloride of Iron.	**IRON** (Ferrous salts). Protosulphate of Iron.	
$Fe_2Cl_3 + 3NH_4HO$ = red-brown..................$Fe_2H_3O_3$ insol. in excess of hydrate of ammonium. insol. in chloride of ammonium.	$Fe_2SO_4 + 2NH_4HO$ = green-white α (becomes red).........$2FeHO$ insol in excess of hydrate of ammonium. sol. in chloride of ammonium β.	I. Hydrates.
hydrate of potassium re-acts in the same manner as hydrate of ammonium on solutions of ferric salts α.	hydrate of potassium re-acts in the same manner as hydrate of ammonium, on solutions of ferrous salts, except that the whole of the metal is precipitated.	II. Hydrates.
$2Fe_2Cl_3 + 3(NH_4)_2S$ = black β..................$(Fe_2)_2S_3$ insol. in excess of sulphide of ammonium, sol. partially, in water, forming a green solution.	$Fe_2SO_4 + (NH_4)_2S$ = black γ..................Fe_2S insol. in excess of sulphide of ammonium, sol partially in water, forming a green solution, and reppt. by sulphide of ammonium.	III. Sulphides.
soluble. green coloration γ.	$3Fe_2SO_4 + 2K_3Cfdy$ = blue..................$2Fe_3Cfdy$ insol. in hydrochloric acid and in water.	IV. Ferri-Cyanides.
$2Fe_2Cl_3 + 3K_4Cfy$ = blue = prussian blue......$(Fe_2)_2Cfy_3$ sol. in nitric acid. sol. in hydrate of potassium.	$3Fe_2SO_4 + 4K_4Cfy$ = white δ..............$2KFe_2Cfy_2$ nitric acid converts it into prussian blue. sol. in hydrate of potassium.	V. Ferro-Cyanides.
soluble.	soluble.	VI. Double Sulphates.
———	———	VII. Silicate.

α The ppt. in this case always contains a portion of the alkaline precipitant, which is not removed by washing.

β This ppt. is only obtained when a solution of ferric chloride is added to sulphide of ammonium, otherwise a ferrous sulphide is thrown down together with sulphur.

Hydrosulphuric acid reduces ferric to ferrous salts, with separation of sulphur if the solution is acid or neutral. If alkaline a ppt. is formed. Both ferrous and ferric salts behave in the same manner with sulphide of ammonium, as the members of Group IV.; hence by some chemists iron is included as a member of that group.

γ This coloration appears to depend upon the presence of a trace of proto- or ferrous salt, and usually acquires a tinge of brown. It is best seen in dilute solutions.

α Only half the iron is precipitated, for the ammonium salt formed combines with part of the unchanged ferrous salt, producing a double salt ($FeCl, NH_4Cl$), which is not decomposed by the excess of hydrate of ammonium present (see Magnesium IV).

β It is clear from the reactions given above that if chloride of ammonium is present in a solution of a ferrous salt, no ppt. will occur on the addition of hydrate of ammonium, hence ferrous salts really belong to Group IV., but in actual analysis the precaution is adopted of peroxidising (by nitric acid) any ferrous salt which may be present in the solution to be tested, before adding hydrate of ammonium as a group test. It is thus converted into a ferric salt, and will behave with the group test, as other members of this group. The desirability of contrasting these two classes of salts of iron will be a sufficient reason for placing them together here.

γ Hydrosulphuric acid does not precipitate acid or neutral solutions of ferrous salts. The sulphide is precipitated from an alkaline solution by this reagent.

δ Turns blue (prussian blue) on exposure to air.

GROUP

METALS WHICH ARE PRECIPITATED FROM SOLUTIONS OF

	MANGANESE. Sulphate of Manganese	ZINC. Sulphate of Zinc.
I. **Sulphide of Ammonium.**	$Mn_2SO_4 + (NH_4)_2S$ α = **flesh-coloured** β............Mn_2S sol. in most acids. *sol. in acetic acid* (if concentrated).	$Zn_2SO_4 + (NH_4)_2S$ α = **white**..........................Zn_2S sol. in most acids *insol. in acetic acid.*
II. **Hydrate of Potassium.**	$Mn_2SO_4 + 2KHO$ γ = **whitish** δ......................$2MnHO$ insol. in excess of hydrate of potassium.	$Zn_2SO_4 + 2KHO$ β = **white** (gelatinous).........$2ZnHO$ sol. in excess of hydrate of potassium.
III. **Carbonate of Ammonium.**	$Mn_2SO_4 + (NH_4)_2CO_3$ = **pink-white**...............Mn_2CO_3 sol. in chloride of ammonium. insol. in excess of carbonate of ammonium.	$5Zn_2SO_4$ + $6H_2O$ } + $2(NH_4)_2CO_3$ = **white** γ............$2(Zn_2CO_3 . 3ZnHO)$ sol. in chloride of ammonium. sol. in excess of carbonate of ammonium.
IV. **Cyanide of Potassium.**	$Mn_2SO_4 + 2KCy$ = **dirty yellow**..............$2MnCy$ sol. in excess of cyanide of potassium ε.	$Zn_2SO_4 + 2KCy$ = **white**..........................$2ZnCy$ sol in excess of cyanide of potassium.
V. **Hydrosulphuric Acid in a solution acidified with Acetic Acid.**	———	$Zn_2SO_4 + 2HC_2H_3O_2 = H_2SO_4 + 2ZnC_2H_3O_2$ acetic acid and $2ZnC_2H_3O_2 + H_2O + H_2S$ = **white**..................Zn_2S, H_2O

Manganese notes:

α Hydrosulphuric acid precipitates alkaline solutions of manganous salts, but not acid or neutral solutions.

β Turns brown on exposure to air; the colour of the above ppt. is extremely characteristic.

γ With hydrate of ammonium only half the manganese is precipitated, the other half forming with the salt already produced by the first half and hydrate of ammonium, a soluble compound which is not re-precipitated by excess of the precipitant. Presence of chloride of ammonium prevents the precipitation by hydrate of ammonium altogether.

δ Turns black on exposure to air.

ε Forms a brown solution.

Zinc notes:

α Hydrosulphuric acid partially precipitates zinc from neutral and entirely from alkaline solutions. In all acid solutions, except acetic acid (see V.), no ppt. occurs.

β The same reaction occurs with hydrate of ammonium as in the case of manganous salts (see Manganese, note γ).

γ The ppt. is a mixed hydrate and carbonate, and since some carbonic anhydride is liberated, which retains a portion of carbonate of zinc in solution, boiling facilitating the escape of carbonic anhydride, precipitation is more complete.

IV.

THEIR SALTS, AS SULPHIDES BY *SULPHIDE OF AMMONIUM*

COBALT. Nitrate of Cobalt.	**NICKEL.** Protosulphate of Nickel.	
$2CoNO_3 + (NH_4)_2S$ α = black .... Co_2S sol. in nitro-hydrochloric acid. *insol. in acetic acid.*	$Ni_2SO_4 + (NH_4)_2S$ α = black β .... Ni_2S sol. in nitro-hydrochloric acid. *insol. in acetic acid.*	I. Sulphides.
$CoNO_3 + KHO$ β = blue γ .... $CoHO$ insol. in excess of hydrate of potassium.	$Ni_2SO_4 + 2KHO$ γ = light-green .... $2NiHO$ insol. in excess of hydrate of potassium.	II. Hydrates.
$10CoNO_3$ + $6H_2O$ } $+ 5(NH_4)_2CO_3$ = peach coloured...... $2(Co_2CO_3 . 3CoHO)$ sol in excess of carbonate of ammonium.	$4Ni_2SO_4$ + $6H_2O$ } $+ 4(NH_4)_2CO_3$ = pale green......... $Ni_2CO_3 . 6NiHO$ sol. in excess of carbonate of ammonium.	III. Carbonates.
$CoNO_3 + KCy$ = brown .... $CoCy$ sol. in excess of cyanide of potassium. add to the solution a few drops of hydrochloric acid, and boil, then add more hydrochloric acid. *the ppt. does not re-form* δ.	$Ni_2SO_4 + 2KCy$ = yellow-green............ $2NiCy$ sol. in excess of cyanide of potassium. add to the solution a few drops of hydrochloric acid, and boil, then add more hydrochloric acid. *the ppt. re-forms* δ.	IV. Cyanides.
——	——	V. Sulphydrate.

α Hydrosulphuric acid partially precipitates black sulphide from neutral solutions of cobaltous salts, but fails to precipitate acid solutions, except the acid radical of the salt be a weak one. If to a solution of a cobaltous salt, treated with acetate of sodium, hydrosulphuric acid is added, complete precipitation occurs, especially on heating. No result occurs however in the presence of free acetic acid.

β Hydrate of ammonium reacts in a similar manner as with manganous and zinc salts: a blue basic salt is however at first thrown down, soluble in excess, forming a red-brown solution.

γ Turns green on exposure to air. On boiling the ppt. becomes converted into the red hydrate, which appears as a dirty-red powder.

δ The addition of hydrochloric acid liberates free hydrocyanic acid reacting with the cyanide of cobalt and cyanide of potassium, cobaltocyanide of potassium being formed at the same time. On boiling cobaltocyanide of potassium ($KCoCy_2$) is converted into cobalti-cyanide ($K_3Co_2Cy_6$)*. This latter containing all the cobalt is not decomposed by dilute acids, hence no ppt. occurs on treating with hydrochloric acid again.

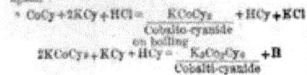

* $CoCy + 2KCy + HCl = \underbrace{KCoCy_2}_{\text{Cobalto-cyanide}} + HCy + KCl$

on boiling

$2KCoCy_2 + KCy + HCy = \underbrace{K_3Co_2Cy_6}_{\text{Cobalti-cyanide}} + H$

α Hydrosulphuric reacts with neutral and acid solutions of salts of nickel, in the same manner as with cobaltous salts. In the case of acetate of nickel however a longer action of the reagent is necessary to throw down a ppt.

β This ppt. is slightly soluble in the reagent, forming a brown solution, from which heat again precipitates the sulphide. Hence a brown colour in the filtrate from the sulphides of this group points out the probable presence of nickel.

γ Hydrate of ammonium produces only a slight ppt. dissolving in a larger quantity of the reagent to a blue liquid, since the hydrate is soluble in the ammonium salt formed at the same time.

δ Here the double cyanide of nickel and potassium is decomposed by dilute hydrochloric acid, and cyanide of nickel is again deposited. But care must be taken that only a small quantity of the acid is added, at most only a few drops, and that in the cold, since excess of acid on boiling converts the cyanide into a soluble salt of nickel.

GROUP

METALS WHICH ARE PRECIPITATED FROM SOLUTIONS OF

	BARIUM. Chloride of Barium.	**STRONTIUM.** Chloride of Strontium.
I. **Carbonate of Ammonium.**	$2BaCl + (NH_4)_2CO_3$ α = white........................Ba_2CO_3 insol. in chloride of ammonium β.	$2SrCl + (NH_4)_2CO_3$ = white........................Sr_2CO_3 insol. in chloride of ammonium.
II. **Sulphate of Calcium.**	$2BaCl + Ca_2SO_4$ = white (immediately) ... Ba_2SO_4 = heavy spar. insoluble.	$2SrCl + Ca_2SO_4$ = white (after some time) ... Sr_2SO_4 = celestine insoluble.
III. **Chromate of Potassium.**	$BaCl + KCrO_2$ = pale yellow....................$BaCrO_2$ sol. in hydrochloric or nitric acid.	soluble α.
IV. **Hydrate of Potassium.**	$BaCl + KHO$ γ = white (bulky)..................$BaHO$ sol. in water, forming baryta water.	$SrCl + KHO$ β = white (flocculent).........$SrHO$ sol. sparingly in water.
V. **Oxalate of Ammonium or Oxalic Acid.**	$2BaCl + (NH_4)_2C_2O_4$ δ = white......... ....$Ba_2C_2O_4$ sol. in acetic acid.	$2SrCl + H_2C_2O_4$ γ = oxalic acid white (immediately)............$Sr_2C_2O_4$ insol. in acetic acid. sol. in boiling chloride of ammonium.
VI. **Ferro-Cyanide of Potassium.**	soluble, unless a *very* concentrated solution of a salt is used.	soluble.
VII. **Phosphate of Sodium.**	$2BaCl + Na_2HPO_4$ = white........................Ba_2HPO_4 sol. in chloride of ammonium, and reppt. by hydrate of ammonium.	$2SrCl + Na_2HPO_4$ = white........................Sr_2HPO_4 sol. in chloride of ammonium, and reppt. by hydrate of ammonium.
VIII. **Hydro-fluo-silicic Acid.**	$3BaCl + H_2Si_2F_6$ = white ε........................$Ba_2Si_2F_6$ insoluble.	soluble.
	α Before adding carbonate of ammonium it is better in the case of all the members of this group to add hydrate of ammonium, to ensure a sufficiently neutral carbonate, and then to heat. β Carbonates of barium, strontium and calcium are slightly soluble in chloride of ammonium. In accurate analysis this entails a further treatment of the filtrate which otherwise would only contain magnesium, with a view to separate the trifling quantities of barium, strontium or calcium which may have been dissolved. γ Hydrate of ammonium produces no ppt. with solutions of barium salts. δ With oxalic acid, the *acid* oxalate of barium is formed ($BaHC_2O_4$), and only after standing for some time. ε Addition of alcohol promotes the formation of the ppt.	α A bright-yellow ppt. ($SrCrO_4$) forms in *concentrated* solutions of a strontium salt, or in *such as are free from acetic acid*. β Hydrate of ammonium produces no ppt. in solutions of a strontium salt. γ Soluble oxalates (oxalate of ammonium) also throws down this ppt.

V.

THEIR SALTS, AS CARBONATES BY *CARBONATE OF AMMONIUM.*

CALCIUM. Chloride of Calcium.	**MAGNESIUM.** Chloride of Magnesium.	
$2CaCl + (NH_4)_2CO_3$ = white................Ca_2CO_3 = calc-spar insol. in chloride of ammonium.	$2MgCl + (NH_4)_2CO_3$ = white (on boiling)..........Mg_2CO_3 α *sol. in chloride of ammonium.*	I. Carbonates.
soluble α.	soluble.	II. Sulphates.
soluble.	soluble.	III. Chromates.
CaCl + KHO β = white (bulky)................CaHO less sol. in *hot* than in *cold* water, forming lime water.	MgCl + KHO β = white (bulky)................MgHO insol. nearly in water.	IV. Hydrates.
2 CaCl + $H_2C_2O_4$ γ = oxalic acid white (after some time)..........$Ca_2C_2O_4$ insol in acetic acid. insol. in chloride of ammonium.	soluble.	V. Oxalates.
CaCl + K_2Cfy = white (on boiling)...........KCaCfy insol. in concentrated hydrochloric acid.	2MgCl + K_2Cfy = yellow-white (on boiling).........Mg_2Cfy γ sol. in hydrochloric acid.	VI. Ferro-Cyanides.
$2CaCl + Na_2HPO_4$ = white.........................Ca_2HPO_4 (?) sol. slightly in chloride of ammonium.	2MgCl + NH_4HO δ } + Na_2HPO_4 = white...............$Mg_2 . NH_4 . PO_4$ insol. except in acids.	VII. Phosphates.
soluble.	soluble.	VIII. Fluo-silicate.

α Sulphate of calcium (Ca_2SO_4) is precipitated from concentrated solutions of a calcium salt, by sulphuric acid. The ppt. forms slowly, but all the calcium may be thrown down if alcohol is added to the solution.

β Hydrate of ammonium produces no ppt. with solutions of a calcium salt.

γ An acid solution of a calcium salt must be neutralised by hydrate of ammonium before adding oxalic acid, as the ppt. is soluble in acids except acetic acid.

Oxalate of ammonium produces the same ppt.

α A soluble salt having the formula [$2Mg . NH_4 . CO_3$] is first formed, heat expels from this carbonate of ammonium [$(NH_4)_2CO_3$], leaving the precipitated carbonate of magnesium.

β Hydrate of ammonium only throws down a portion of the magnesium. The other portion remains in solution, as a double chloride of magnesium and ammonium. The latter is formed as follows, decomposition of the magnesium salt (here chloride of magnesium) first takes place; the ammonium salt then formed unites with another equivalent of magnesium salt to form the double salt. This salt is undecomposable.

$$2MgCl + NH_4HO = MgCl . NH_4Cl + MgHO.$$

γ If an ammonium salt is present, such as chloride of ammonium (which hastens precipitation), the formula becomes $Mg . NH_4Cfy$.

δ Hydrate of ammonium is added in order to ensure complete precipitation of the magnesium. If however it is too strong, a ppt. is formed with it and phosphate of sodium. In dilute solutions of a magnesium salt, the double salt forms only after some time. This reaction is an important test for magnesium.

* Phosphate of sodium should be added first, and then hydrate of ammonium.

GROUP

METALS WHICH ARE *NOT* PRECIPITATED FROM SOLUTIONS OF

	POTASSIUM. Chloride of Potassium.	SODIUM. Chloride of Sodium.
I. **Hydro-Chloro-platinic Acid.**	$KCl + HCl\,\alpha + HPtCl_3 =$ yellow (crystalline)............$KPtCl_3$ insol. in alcohol and in acids.	soluble.
II. **Tartaric Acid.**	$KCl + H_2C_4H_4O_6\,\beta =$ white........................$KHC_4H_4O_6$ sol. in hydrate of potassium.	soluble.
III. **Hydrate of Potassium, and Hydro-chloric Acid.**	——— γ	——— α
IV. **Carbonate of Ammonium.**	———	———
V. **Phosphate of Sodium.**	———	———
VI. **Metantimon-iate of Potassium.**	———	$2NaCl + K_2H_2Sb_2O_7\,\beta =$ white........................$Na_2H_2Sb_2O_7$ insol. in cold water. sol. slightly in boiling water.

α Hydrochloric acid is added in order to render the solution of potassium sufficiently acid. No ppt. is produced in alkaline solutions.

When only a small quantity of potassium is present in the solution to be tested, add hydrochloric acid and hydrochloroplatinic acid, evaporate to dryness and digest with alcohol. The chloroplatinate remains as a yellow residue.

β Excess of the precipitant must be added, or the soluble *neutral* tartrate ($K_2C_4H_4O_6$) will be formed. Vigorous shaking promotes the formation of the ppt.

γ Chloride of potassium is not affected like chloride of ammonium in this experiment, because, when decomposed by hydrate of potassium it is not volatile.

α Chloride of sodium is not affected here, for the same reason as that given in the case of potassium.

β Care should be observed to prepare this solution just before use, since the metantimoniate is prone to pass into antimoniate. $K_2H_2Sb_2O_7 - H_2O = 2KSbO_3$.

The solution of sodium salt must be neutral or slightly alkaline, as free acid separates antimonic acid from the reagent.

VI.

THEIR SALTS BY *REAGENTS WHICH PRECIPITATE THE OTHER GROUPS.*

AMMONIUM. Chloride of Ammonium.	**LITHIUM.** Chloride of Lithium.	
$NH_4Cl + HPtCl_3$ = yellow (crystalline)............NH_4PtCl_3 insol. in cold or boiling water.	soluble.	I. Chloro-platinates.
$NH_4Cl + H_2C_4H_4O_6$ = white α (crystalline).........$NH_4HC_4H_4O_6$ sol. in hydrate of ammonium.	soluble.	II. Acid Tartrates.
$NH_4Cl + KHO$ = gas β..............................NH_3 a glass rod dipped in *dilute* γ hydrochloric acid, and applied to the mouth of the test tube, gives rise to *white fumes.* $NH_3 + HCl$ = white fumes..................NH_4Cl	———	III. Decomposition.
———	$2LiCl + (NH_4)_2CO_3$ = white (crystalline)............Li_2CO_3 insoluble α.	IV. Carbonate.
———	$LiCl + Na_2HPO_4$ = white (especially on heating)...$LiNaHPO_4$(?) insoluble α.	V. Phosphate.
———	———	VI. Acid Metantimoniate.
α This ppt. is only produced in very concentrated solutions of ammonium salts. β Gaseous ammonia is readily detected by its pungent odour. γ Concentrated acids give rise themselves to fumes on exposure to air, hence dilute acid must be employed in this experiment.	α The insolubility of these salts forms a remarkable exception to the characters of the salts of this group generally.	

PART II.

ACID RADICALS.

CLASSIFICATION OF ACID RADICALS AND THEIR CORRESPONDING HYDROGEN SALTS.

Group I.

Acid radicals, which are precipitated from their *neutral* solutions, by *Chloride of Barium.*

Section I.

Barium compounds, *insoluble in Hydrochloric Acid.*

Sulphuric radical (sulphuric acid).
Silicofluorine (hydrofluosilicic acid).

Section II.

Barium compounds, *soluble in Hydrochloric Acid,* with decomposition.

Carbonic radical (carbonic acid).
Silicic radical (silicic acid).
[Hydrosulphuric acid].

Section III.

Barium compounds, *soluble in Hydrochloric Acid,* without decomposition.

Phosphoric radical (phosphoric acid).
Boracic radical (boracic acid).
Oxalic radical (oxalic acid).
Fluorine (hydrofluoric acid).

Section IV.

Decomposed, in acid solution, by *Hydrosulphuric Acid.*

Chromic radical (chromic acid).
Sulphurous radical (sulphurous acid).
Hyposulphurous radical (hyposulphurous acid).
Iodic radical (iodic acid).
[Arsenious and Arsenic acids].

Group II.

Acid radicals, which are precipitated from their *acidified* (nitric acid) solutions, by *Nitrate of Silver.*

Section I.

Argentic salts, soluble in *Hydrate of Ammonium.*

Chlorine (hydrochloric acid).
Bromine (hydrobromic acid).
Cyanogen (hydrocyanic acid).
Ferricyanogen (hydro-ferri-cyanic acid).

Section II.

Argentic salts, insoluble in *Hydrate of Ammonium.*

Iodine (hydriodic acid).
Sulphur (hydrosulphuric acid).
Ferrocyanogen (hydro-ferro-cyanic acid).
Sulphocyanogen (hydro-sulpho-cyanic acid).

[The following acid radicals are precipitated from their *neutral* solutions by *Nitrate of Silver*: silicic, phosphoric, boracic, oxalic, chromic, chlorine, bromine, iodine, cyanogen, sulphur, (arsenious and arsenic radicals).]

Group III.

Acid radicals, which are *not precipitated* from their solutions by *Salts of Barium or Silver.*

Nitric radical (nitric acid).
Chloric radical (chloric acid).

ORGANIC ACIDS.

Group I.

Acid radicals, which are precipitated from their solutions by *Chloride of Calcium.*

Tartaric radical (tartaric acid).
Citric radical (citric acid).

Group II.

Acid radicals, which are precipitated from their *neutral* solutions by *Ferric Chloride.*

Succinic radical (succinic acid).
Benzoic radical (benzoic acid).
Tannic radical (tannic acid).
Gallic radical (gallic acid).

Group III.

Acid radicals, which are *not precipitated* from their solutions by *Chloride of Calcium or Ferric Chloride.*

Acetic radical (acetic acid).
Uric radical (uric acid).

GROUP I. Section i.

*ACIDS, WHICH ARE PRECIPITATED FROM SOLUTIONS OF THEIR SALTS BY *CHLORIDE OF BARIUM.*

THE PRECIPITATE *INSOLUBLE IN HYDROCHLORIC ACID.*

	***SULPHURIC ACID** (H_2SO_4). Sulphate of Magnesium.	**HYDROFLUOSILICIC ACID** ($H_3Si_2F_9$). Fluosilicic Acid.	
I. **Chloride of Barium.**	$Mg_2SO_4 + 2BaCl$ = white α .... Ba_2SO_4 insol. in hydrochloric acid.	$H_3Si_2F_9 + 3BaCl$ = white (crystalline) α .... $Ba_3Si_2F_9$ insol. in hydrochloric acid.	I. Barium Salts.
II. **Chloride of Calcium.**	$Mg_2SO_4 + 2CaCl$ = white .... Ca_2SO_4 insol. in acetic acid. sol. in much water and hot hydrochloric acid.	soluble.	II. Calcium Salts.
III. **Heated with Conc^d H_2SO_4.**	——	**Dense fumes β.** Test. Conduct the experiment in a platinum crucible, covered by a plate of glass; *the fumes etch the glass.*	III. Decomposition.
IV. **Fused on Charcoal with Na_2CO_3.**	**Reduction to sulphide.** Tests. (*a*) Add HCl to the reduced mass; *odour of hydrosulphuric acid gas.* (*b*) Add HCl to the mass on a clean silver surface; *black stain on the silver β.*	——	IV. Blowpipe reaction.
	α If the solution of sulphate contains an excess of free hydrochloric or nitric acids, the chloride or nitrate of barium insoluble in strongly acid solutions will be precipitated, and thus interfere with the value of the above reaction. If, however, the solution is largely diluted with water before adding chloride of barium, these salts, should they be formed, will be immediately dissolved. The dilution of the solution should never be neglected in testing for sulphuric acid with chloride of barium. The addition of a *little* hydrochloric acid should also precede that of chloride of barium, as this counteracts the influence of certain citrates in case they may be present. β The black stain is due to the formation of sulphide of silver (Ag_2S). Any compound of sulphur will exhibit this reaction, and therefore it cannot be relied upon as a characteristic test for sulphuric acid.	α Addition of alcohol hastens and completes the precipitation: this reaction furnishes also a characteristic test for barium (see Barium). β The fumes are due to the formation of terfluoride of silicon (SiF_3), thus— $H_3Si_2F_9 = 2SiF_3 + 3HF$ The etching results from the action of the liberated hydrofluoric acid (HF) upon the silica of the glass. Thus— $6HF + Si_2O_3 = 2SiF_3 + 3H_2O$ (See hydrofluoric acid, p. 47.) Hydrofluosilicic acid also decomposes when exposed to the air, or when evaporated, and if in glass vessels the glass is attacked by the liberated hydrofluoric acid.	

* The names of the hydrogen-salts of acid-radicals (so-called "acids") have been placed at the head of the columns, in the following pages, on the ground of expediency. Conventional terms have been retained, in many instances, for the same reason.

GROUP I. Section II.

ACIDS, WHICH ARE PRECIPITATED FROM SOLUTIONS OF THEIR SALTS BY *CHLORIDE OF BARIUM.*

THE PRECIPITATE SOLUBLE IN HYDROCHLORIC ACID *WITH DECOMPOSITION*

	CARBONIC ACID (H_2CO_3). Carbonate of Sodium.	**SILICIC ACID** ($HSiO_2$) α. Silicate of Potassium.	
I. **Chloride of Barium.**	$Na_2CO_3 + 2BaCl$ = white .... Ba_2CO_3 sol. in hydrochloric acid, if heated gently to expel carbonic anhydride; not reppt. by hydrate of ammonium.	$KSiO_2 + BaCl$ = white .... $BaSiO_2$ sol. in hydrochloric acid. insol. in water β.	I. Barium Salts.
II. **Chloride of Calcium.**	$Na_2CO_3 + 2CaCl$ = white .... Ca_2CO_3 sol. in acetic acid.	$KSiO_2 + CaCl$ = white .... $CaSiO_2$ sol. in hydrochloric and most acids. insol. in water.	II. Calcium Salts.
III. **Nitrate of Silver.**	$Na_2CO_3 + 2AgNO_3$ = white .... Ag_2CO_3 sol. in nitric acid and hydrate of ammonium.	$KSiO_2 + AgNO_3$ = white .... $AgSiO_2$ sol. in acids. insol. in water.	III. Argentic Salts.
IV. **Heated with Hydrochloric Acid.**	**Effervescence** α. (evolution of carbonic anhydride) $Na_2CO_3 + 2HCl = H_2CO_3 + 2NaCl$ and $H_2CO_3 = CO_2 + H_2O$ Test. Conduct the gas into lime-water or baryta-water; $2CaHO + CO_2 = Ca_2CO_3 + H_2O$. *white ppt.* β.	i. HCl + *concentrated* solution of alkaline silicate γ. **White ppt.** $KSiO_2 + HCl = HSiO_2 + KCl$ (ppt. = gelatinous or hydrated silicic acid). ii. HCl + *dilute* solution of alkaline silicate. **No ppt.** (silicic acid remains in solution δ). Test. Evaporate the solution to dryness; add HCl and water to residue; evaporate again; *white powder separates* (silicic anhydride ε).	IV. Decompositions.
V. **Fused on Platinum Wire with Na_2CO_3.**	——	**Transparent bead.** (cold.) (characteristic).	V. Blowpipe reaction.

α This reaction takes place even with the weakest acids*.

This does not occur however with hydrocyanic or hydrosulphuric acid†.

Substances tested for carbonic acid by this method, if in the solid state, should first be heated with water, as otherwise air-bubbles will be disengaged on adding the acid, thus simulating the escape of carbonic anhydride.

β This ppt. dissolves if excess of carbonic anhydride is passed into the lime-water, from formation of the soluble acid carbonate of calcium.

$$CaHO + CO_2 = \underbrace{CaHCO_3}_{\text{acid salt}}$$

* The acid must be added in excess.

† Salts of other acid radicals (cyanates, sulphides, &c.) are decomposed by acids with effervescence, but in such cases the escaping gas is sufficiently recognised by characteristic odour and other reactions.

α The formula here given for silicic acid supposes the atomic weight of silicon to be 22·0, the anhydride then becomes Si_2O_4.

β The silicates of potassium and sodium are the only ones soluble in water.

γ Hydrochloric acid should be added drop by drop, and the solution stirred. The great majority of silicates are not decomposable by acids (hydrofluoric excepted). They may however be rendered decomposable by fusing them with three or four times their weight of the mixed carbonates of potassium and sodium. The fused mass is readily soluble in hydrochloric acid, and the silica may be separated by evaporating to dryness and treating with hydrochloric acid, as in IV.

To detect alkaline metals in silicates, hydrate of barium must be substituted for the mixed carbonates.

Before fusion, or treating with acids, the silicate should be reduced to an extremely fine pulverulent form in a mortar. Nitric acid is used in the case of silicates containing lead.

δ Doveri states that solution does not occur, but is only apparent from the extreme transparency of the precipitated hydrate.

Silicic acid is only soluble in acids at *the moment of separation* from its solutions, and then appears to enter into combination with the acid (Graham).

ε Silicic anhydride is absolutely soluble; this is often termed insoluble silica, silicic acid being termed soluble silica. This solubility occurring however only under the condition stated in the foregoing note.

GROUP I.

ACIDS, WHICH ARE PRECIPITATED FROM SOLUTIONS ... THE PRECIPITATE IS SOLUBLE IN HYDRO-

	PHOSPHORIC ACID (H_3PO_4)α. Phosphate of Sodium.	**BORACIC ACID** ($HBoO_2$)α. Bi-borate of Sodium (Borax).
I. **Chloride of Barium.**	$Na_2HPO_4 + 2BaCl$ = white..............................Ba_2HPO_4 sol. in hydrochloric acid, reppt. by hydrate of ammonium. insol. in water.	$Na_2Bo_4O_7 + 2BaCl$ = white..............................$Ba_2Bo_4O_7$ sol. in hydrochloric acid, not reppt. by hydrate of ammoniumβ. sol. sparingly in water.
II. **Chloride of Calcium.**	$Na_2HPO_4 + 2CaCl$ = white..............................Ca_2HPO_4 sol. in hydrochloric or acetic acid, reppt. by hydrate of ammonium.	$Na_2Bo_4O_7 + 2CaCl$ = white..............................$Ca_2Bo_4O_7$ sol. in hydrochloric or acetic acid. sol. in ammonium salts.
III. **Nitrate of Silver.**	$Na_2HPO_4 + 3AgNO_3$ = yellow..............................Ag_3PO_4 sol. in nitric or acetic acid and in hydrate of ammonium.	$Na_2Bo_4O_7 + 2AgNO_3 = Bo_2O_3 +$ whiteγ..............................$2AgBoO_2$ sol. in nitric acid and in hydrate of ammonium.
IV. **Ferric Chloride.**	$Na_2HPO_4 + Fe_2Cl_3$ = white-yellowβ..............$(Fe_2)_2PO_4$(?) insol. in acetic acid. sol. in other acids.	——
V. **Molybdate of Ammonium.**	$Na_2HPO_4 + \gamma HNO_3 + NH_4MoO_2$ = yellow (formula unknown) .... compound of phosphoric and molybdic acids with ammonium insol. in dilute acids. sol. in alkalies and their carbonates.	——
VI. **Heated with Concentrated H_2SO_4, alone and with other reagents.**	——	Shining scales crystallise out on cooling. Test. Treat the scales with alcohol and ignite; *burn with a green flame*δ.

α Phosphoric anhydride (P_2O_5) when dissolved in water may form three different hydrogen salts according to the conditions under which the solution takes place. These are

$P_2O_5 + H_2O = 2HPO_3$ meta- or monobasic phosphoric acid.

$P_2O_5 + 2H_2O = H_4P_2O_7$ pyro- or bibasic phosphoric acid.

$P_2O_5 + 2H_2O = 3H_3PO_4$ ortho- or tribasic phosphoric acid.

The latter is only of sufficient importance to be considered here.

β The formula of this salt is uncertain. The ppt. forms in neutral, slightly alkaline, or solutions containing *free acetic acid*. If the solution contains hydrochloric or any other mineral acid, this may be replaced by acetic acid, by addition of acetate of potassium or sodium in excess—presence of tartaric acid and certain organic matters prevent the reaction.

γ The addition of some hydrate of ammonium should precede that of nitric acid. Care should be taken not to operate with excess of phosphate.

α This may be considered as representing the type of the monoborates. The anhydride combined with a monoborate may be taken as the type of the biborates—thus $Bo_2O_3 + 2HBoO_2 = H_2Bo_4O_7$.

A solution of boracic acid, or of an alkaline borate, *turns turmeric paper red*. (This reaction should be contrasted with the effect of other acids.)

β Not reppt. only if excess of hydrochloric acid is present.

γ Only precipitated from concentrated solutions. Before concentrating dilute solutions of boracic acid, the latter must be combined with an alkali, otherwise a large portion would volatilize on evaporation.

δ The green coloration is most apparent at the edge of the flame when viewed against a dark background.

Salts of copper also manifest this characteristic colour. The presence of metallic chlorides gives rise to the formation of chloride of ethyl, which also produces a green coloration of the flame.

SECTION III.

OF THEIR SALTS, BY *CHLORIDE OF BARIUM.*

CHLORIC ACID, *WITHOUT DECOMPOSITION.*

OXALIC ACID ($H_2C_2O_4$). Oxalate of Ammonium.	**HYDROFLUORIC ACID** (HF). Fluoride of Potassium.	
$(NH_4)_2C_2O_4\alpha + 2BaCl =$ white .... $Ba_2C_2O_4$ sol. in hydrochloric acid, reppt. by hydrate of ammonium. sol. sparingly in water.	$KF + BaCl =$ white .... BaF sol. in hydrochloric acid, reppt. by hydrate of ammonium. insol. in water.	I. Barium Salts.
$(NH_4)_2C_2O_4 + 2CaCl =$ white β .... $Ca_2C_2O_4$ sol. in hydrochloric acid. insol. in acetic acid.	$KF + CaCl =$ white α = fluor spar. .... CaF sol. sparingly in hydrochloric acid, reppt. by hydrate of ammonium. insol. in acetic acid.	II. Calcium Salts.
$(NH_4)_2C_2O_4 + 2AgNO_3 =$ white .... $Ag_2C_2O_4$ sol. in nitric acid, and hydrate of ammonium	—	III. Argentic Salts.
—	—	IV. Ferric Salt.
—	—	V. Molybdate, &c.
i. H_2SO_4 + oxalate. **Effervescence.** (evolution of carbonic oxide and anhydride.) $H_2C_2O_4 + H_2SO_4 = H_2SO_4 . H_2O + CO + CO_2$ *No blackening of the compound* γ. Test. Ignite the evolved gases; *carbonic oxide burns with blue flame.* ii. $H_2SO_4 + Mn_2O_3$ + an oxalate. **Effervescence.** (evolution of carbonic anhydride only.)	i. H_2SO_4 + fluoride. **Dense fumes.** (in moist air.) $2KF + H_2SO_4 = 2HF + K_2SO_4$ fumes Tests. (*a*) *Pungent odour.* (*b*) *Corrosive action on glass* γ. $6HF + Si_2O_3 = 2SiF_3 + 3H_2O$. ii. H_2SO_4 + Sand (Si_2O_3) + fluoride (the products being conveyed into water δ). **Deposit of silica.** (gelatinous ppt.) $3SiF_3 + 2H_2O = HSiO_2 + H_4Si_2F_6$. ppt. Test. Add chloride of barium to filtrate from $HSiO_2$ ppt.; *ppt. of fluosilicate of barium* ($Ba_2Si_2F_6$).	VI. Decompositions.

α If oxalic acid is used, the acid oxalate is formed ($BaHC_2O_4$); the reactions with solvents are the same in each case.

β In dilute solutions the ppt. appears only after some time.

Oxalates of chromium and iron are not precipitated by chloride of calcium, as they form soluble double salts with oxalate of calcium.

γ This reaction is characteristic, since most complex acid radicals (ferrocyanide excepted) blacken as well as yield carbonic oxide.

α Addition of hydrate of ammonium promotes precipitation, which usually without this precaution is hardly perceptible.

β This reaction cannot be observed if the experiment is conducted in a glass vessel, or if silicon is present in any form. If silica is present, or the compound is not decomposable by sulphuric acid, the following plan must be adopted:—"Fuse the substance with carbonate of sodium and potassium—dissolve the fused mass in water—precipitate the silica by carbonate of ammonium and filter it off—neutralise the filtrate with acetic acid, and add chloride of calcium."

γ The etching property of hydrofluoric acid depends on the remarkable affinity of fluorine for silicon. The finely divided fluoride should be placed in a platinum crucible, concentrated sulphuric acid poured on it, and the whole covered with a glass coated with wax through which a device has been scratched. The wax may afterwards be removed by turpentine, when the etching will be plainly visible—$6CaF + 4H_2SO_4 + Si_2O_3 = 2HF + 2SiF_3 + 4Ca_2SO_4 + 3H_2O$. Hydrofluoric Acid is first formed, and this reacts upon the Silica of the glass. (See Hydrofluo-silicic Acid, p. 44, note β.)

δ This experiment may be performed in a test-tube furnished with a cork and bent tube, the latter dipping into mercury, covered with a layer of water. The apparatus must be perfectly dry for obvious reasons. To detect very small quantities of hydrofluoric acid, the delivery tube may be wetted internally, when the deposit of silica in minute quantity will be discovered.

GROUP I.

ACIDS, WHICH ARE PRECIPITATED FROM ALKALINE OR NEUTRAL ACID SOLUTIONS OF THEIR SALTS

	CHROMIC ACID ($HCrO_2$)*. Chromate of Potassium.	**SULPHUROUS ACID** (H_2SO_3). Sulphite of Ammonium.
I. **Chloride of Barium.**	$KCrO_2 + BaCl$ = **pale yellow**............... $BaCrO_2$ sol. in hydrochloric and nitric acid, reppt. from solution in the former by hydrate of ammonium. insol. in water.	$(NH_4)_2SO_3 + 2BaCl$ = **white**............... Ba_2SO_3 sol. in hydrochloric acid α, not reppt. by hydrate of ammonium if boiled to expel the sulphurous acid.
II. **Nitrate of Silver.**	$KCrO_2 + AgNO_3$ = **crimson** (crystalline) .... $AgCrO_2$ sol. in nitric acid and hydrate of ammonium.	$(NH_4)_2SO_3 + 2AgNO_3$ = **white** (granular)............... Ag_2SO_3 becomes dark grey when boiled β. sol. in hydrate of ammonium.
III. **Acetate of Lead.**	$KCrO_2 + PbC_2H_3O_2$ = **yellow**............... $PbCrO_2$ insol. in water and chloride of ammonium. sol. in nitric acid and hydrate of potassium.	$(NH_4)_2SO_3 + 2PbC_2H_3O_2$ = **white**............... Pb_2SO_3 insol. in water.
IV. **Hydrosulphuric Acid Gas, passed through a solution.**	i. Acid solution (HCl) of a chromate. $4KCrO_2 + 10HCl + 3H_2S = 2Cr_2Cl_3 +$ **white-yellow** α............... 3S (ppt.). ii. Neutral or alkaline solution of a chromate. $4KCrO_2 + 5H_2S$ = 3S (ppt.) + **green** β............... $2Cr_2H_3O_3$	$5H_2SO_3$ γ $+ 5H_2S = H_2S_5O_6 +$ **white-yellow**............... 5S (ppt.)
Hydrochloric Acid.	The reaction is similar to that exhibited by sulphuric acid and a chromate. Chlorine is evolved.	$(NH_4)_2SO_3 + 2HCl = SO_2 + H_2O + 2NH_4Cl$ *no precipitation of sulphur* δ.
Zinc and Hydrochloric Acid.	——	**Evolution of H_2S gas.** Test. Contact with paper moistened with acetate of lead; *paper is blackened.*
Heated with Concd H_2SO_4.	**Evolution of Oxygen** γ. Test. Apply a piece of kindled wood to escaping gas; *bursts into flame.*	**Effervescence.** (evolution of sulphurous anhydride.) Test. *Odour of burning sulphur.*

α The sesquichloride of chromium has a green colour: this together with the pale yellow ppt. of sulphur is characteristic of chromic salts, and betrays their presence in the systematic course of analysis for bases.

β In this case a green ppt. of chromic hydrate accompanies the sulphur ppt.

γ $K_2Cr_2O_7 + 4H_2SO_4 = K_2SO_4 + (Cr_2)(SO_4)_3 + 4H_2O + O_3$
bichromate

* The hydrogen salt has not been isolated. The anhydride (Cr_2O_3) is obtained in the form of dark crimson crystals, from an aqueous solution of chromic acid.

α If this hydrochloric acid solution is boiled with nitric acid, the sulphite is converted into sulphate; this being an insoluble salt is precipitated.

β The white sulphite is decomposed into sulphuric acid and metallic silver.

γ This decomposition only occurs in the case of free sulphurous acid. If the experiment is performed with a sulphite, the latter must be previously treated with hydrochloric acid, as in the next reaction.

δ This reaction with hydrochloric acid, *without precipitation of sulphur*, affords a means of distinguishing sulphurous from hyposulphurous acid.

SECTION IV.

SOLUTIONS OF THEIR SALTS, BY *CHLORIDE OF BARIUM.*
DECOMPOSED BY HYDROSULPHURIC ACID.

HYPOSULPHUROUS ACID ($H_2S_2O_3$). Hyposulphite of Sodium.	**IODIC ACID** (HIO_3). Iodate of Potassium.	
$Na_2S_2O_3 + 2BaCl$ = white.............................$Ba_2S_2O_3$ sol. in hydrochloric acid α. sol in boiling water.	$KIO_3 + BaCl$ = white (granular)..........$BaIO_3$ sol. in nitric acid. sol. sparingly in water.	I. Barium Salt.
$Na_2S_2O_3 + 2AgNO_3$ = white.............................$Ag_2S_2O_3$ becomes yellow, red, and finally black β. sol. in excess of hyposulphite of sodium.	$KIO_3 + AgNO_3$ = white.........................$AgIO_3$ sol. in hydrate of ammonium. sol. sparingly in nitric acid.	II. Argentic Salt.
$Na_2S_2O_3 + 2PbC_2H_3O_2$ = white (becomes black when cool) γ...$Pb_2S_2O_3$ sol. in alkaline hyposulphites.	$KIO_3 + PbC_2H_3O_2$ = white.........................$PbIO_3$ sol. sparingly in water and nitric acid.	III. Plumbic Salt.
This acid is introduced as a member of Section I. because since in practice the members of this section are detected in testing for bases, and the addition of hydrosulphuric acid is preceded by hydrochloric acid, the mere addition of the latter is sufficient to decompose the solution of a hyposulphite with separation of sulphur.	$2KIO_3 + 3H_2S$ = H_2SO_4 + white yellow α............2S (ppt.)	IV. Decompositions.
$Na_2S_2O_3 + 2HCl$ = SO_2 + white-yellow δ................S (ppt.)	——	
Reaction the same as in the case of sulphurous acid.	——	
Evolution of sulphurous anhydride.	——	

α Nitric acid converts the ppt. into sulphate, which is insoluble.
β The change in colour results from decomposition of the sulphite, and formation of the black sulphide of silver.
γ Nitric acid converts this ppt. into sulphate.
δ Hyposulphurous acid is first set free, and rapidly decomposed into sulphurous acid and sulphur. The sulphurous acid further decomposes into water and sulphurous anhydride, the presence of the latter being revealed by its characteristic odour.

α The decomposition of iodic acid by hydrosulphuric acid is attended with evolution of free iodine, and subsequent formation of hydriodic acid.

GROUP II.

ACIDS, WHICH ARE PRECIPITATED FROM SOLUTIONS · THE PRECIPITATE *SOLUBLE* IN

	HYDROCHLORIC ACID (HCl). Chloride of Sodium.	**HYDROBROMIC ACID** (HBr). Bromide of Potassium.
I. **Nitrate of Silver.**	$NaCl + AgNO_3$ = **white** (curdy) α..................AgCl sol. in hydrate of ammonium, reppt. by nitric acid. insol. in *dilute* nitric acid. fuses without decomposition on being heated.	$KBr + AgNO_3$ = **white-yellow** (curdy) ....AgBr sol. in hydrate of ammonium, reppt. by nitric acid. insol. in *dilute* nitric acid. fuses without decomposition on being heated.
II. **Ferrous Sulphate and Ferric Chloride.**	—	—
III. **Sulphide of Ammonium.**	—	—
IV. **Treated with Conc^d Sulphuric Acid, alone and with other reagents.**	i. Solid chloride + H_2SO_4 β. **Fumes of HCl gas.** (effervescence.) $2NaCl + H_2SO_4 = 2HCl + Na_2SO_4$ ii. Solid chloride + H_2SO_4 + a chromate γ. **Red fumes,** condensing to a *blood-red liquid* (chloro-chromic acid) δ. $NaCl + H_2SO_4 + KCrO_2 = KNaSO_4 + H_2O +$ red liquid.......................CrOCl Tests. (a) Add hydrate of ammonium to liquid; *colour becomes yellow.* $CrOCl + 2NH_4HO = NH_4Cl + NH_4.CrO_2 + H_2O$ (b) Add slight excess of acetic acid to the yellow solution: then nitrate of silver; *crimson ppt.* $NH_4CrO_2 + AgNO_3 = AgCrO_2 + NH_4NO_3$ iii. Solid chloride + $H_2SO_4 + Mn_2O_2$. **Evolution of chlorine gas.** $2NaCl + H_2SO_4 = 2HCl + Na_2SO_4$ and $4HCl + Mn_2O_2 = Cl_2 + 2MnCl + 2H_2O$. Tests. *Odour. Colour. Bleaches moist vegetable colours.*	i. Solid bromide + H_2SO_4 α. **Evolution of bromine vapour** β. $2KBr + H_2SO_4 = 2HBr + K_2SO_4$ γ. and $2HBr + H_2SO_4 = 2Br + H_2SO_3 + H_2O$. Tests. (a) *Pungent odour. Red-brown colour.* (b) Contact with starch-paste on a glass rod; *yellow or orange colour.* ii. Solid bromide + H_2SO_4 + a chromate. **Evolution of bromine vapour** δ. (condenses to red-brown liquid.) $2KBr + H_2SO_4 + KCrO_2 = 2HBr + K_2SO_4 + KCrO_2$ and $2HBr + H_2SO_4 = 2Br + H_2SO_3 + H_2O$ Tests. (a) Add hydrate of ammonium to liquid; *becomes colourless.* (b) Add nitrate of silver to the colourless solution; *white ppt.* iii. Solid bromide + $H_2SO_4 + Mn_2O_2$. **Evolution of bromine vapour.** [the reaction is similar with hydrochloric acid.] Test, as above, i.

α This ppt. becomes violet on exposure to light *.
β The chlorides of mercury, silver, lead and tin resist decomposition by sulphuric acid.
γ (Solid chlorides in presence of a nitrate, when heated with concentrated sulphuric acid, evolve chlorine and nitrous fumes.)
The chloride in this experiment should be perfectly dry, as also the apparatus in which it is performed. The gas may be conducted from the test tube in which generated, to another surrounded with water, to condense it.
Chloro-chromic acid is readily decomposed by water.
δ The formula is sometimes written
$CrCl_3 . Cr_2O_3 = 3(CrOCl)$.

* Most of the acids of Group I. are also precipitated by nitrate of silver, but the silver salts of Group I. are sufficiently distinguished from those of Group II. by the solubility of the former in dilute nitric acid.

α With dilute sulphuric acid, hydrobromic acid gas is evolved.
β Chlorine-water (not in excess) also liberates bromine when added to a solution of a bromide. The bromine dissolves in the water, rendering it yellow or orange according to quantity present. If this solution is agitated with some ether, the latter dissolves out the bromine from the water, and on standing rises with it to the surface of the liquid, as a red-brown layer. This layer, if decanted and agitated with hydrate of potassium, is decolorised from formation of bromate of potassium. On evaporation and ignition the latter is converted into bromide of potassium. (This reaction will not distinguish between bromine and iodine.)
γ If sulphuric acid is added to a bromide, in the cold, this reaction only takes place; hydrobromic acid being liberated, heat causes the further decomposition.
δ No bromo-chromic acid is known.

SECTION I.

OF THEIR SALTS, BY *NITRATE OF SILVER.*
HYDRATE OF AMMONIUM.

HYDROCYANIC ACID (HCy). Cyanide of Potassium.	**HYDRO-FERRI-CYANIC ACID** ($H_3Fe_2Cy_6 = H_3Cfdy$). Ferricyanide of Potassium.	
$KCy + AgNO_3$ = white (curdy) .... $AgCy$ sol. in hydrate of ammonium, and concd nitric acid. insol. in *dilute* nitric acid. decomposed on ignition with separation of metallic silver and cyanogen.	$K_3Cfdy + 3AgNO_3$ = orange .... Ag_3Cfdy sol. in hydrate of ammonium. insol. in nitric acid. decomposed by heat with separation of metallic silver.	**I.** **Argentic Salts.**
A mixture of ferrous sulphate and ferric chloride α. $6KCy + Fe_2SO_4 = 2(K_2FeCy_3) + K_2SO_4$ and $3(K_2FeCy_3) + 2Fe_2Cl_3$ = blue = prussian blue.....$(Fe_2)_2(FeCy_3)_3$	$2K_3Cfdy + 3Fe_2SO_4$ = blue .... $2Fe_3Cfdy$ insol. in water and hydrochloric acid. the ferric salt is soluble α.	**II.** **Ferrous and Ferric Salts.**
$KCy + HCl = HCy + KCl$ β and $HCy + NH_4HS + S = NH_4CyS + H_2S$ The sulphocyanide of ammonium is treated with ferric chloride. $3NH_4CyS + Fe_2Cl_3$ = blood-red solution...... $Fe_2(CyS)_3 + 3NH_4Cl$	—	**III.** **Ferric Sulpho-cyanide.**
i. Solid cyanide + H_2SO_4 γ. **Evolution of carbonic oxide.** $\left.\begin{matrix} 2KCN + 2H_2O \\ + \\ 2H_2SO_4 \end{matrix}\right\} = \left\{\begin{matrix} (NH_4)_2SO_4 \\ + \\ \underset{gas}{\underline{2CO}} + K_2SO_4 \end{matrix}\right.$ Test. Ignite the gas; *burns with blue flame.* [With *dilute* sulphuric acid or hydrochloric acid, cyanides evolve hydrocyanic acid gas (see above III): much caution should be exercised in performing this experiment from the poisonous nature of the vapours. They burn with a blue flame.]	Ferricyanides behave like cyanides with concentrated sulphuric acid and with dilute sulphuric or hydrochloric acid.	**IV.** **Decomposition.**

α The addition of hydrate of potassium must precede that of the mixed salts of iron (since the reaction does not occur in presence of free hydrocyanic acid). And since the alkali itself forms a dense ppt. of ferrous and ferric hydrate, which obscures the prussian blue ppt., it is necessary further to dissolve the hydrate ppt. in some dilute hydrochloric acid.

β This experiment should be performed in the following manner—place the cyanide in a watch-glass; add a few drops of hydrochloric acid; invert another watch-glass over the first, the interior being moistened with sulphide of ammonium containing an excess of sulphur. After the sulphocyanide of ammonium is formed, evaporate all remaining sulphide of ammonium from the upper watch-glass in a water-bath, then add the ferric chloride to the residue.

γ To detect hydrocyanic acid in cyanide of mercury neither this method nor the preceding one will apply. The mercury must first be precipitated as sulphide by hydrosulphuric acid.

α The solution of a ferricyanide after adding ferric chloride assumes a green tint. This coloration appears to depend on the invariable presence in practice of a trace of a ferrous salt.

GROUP II.

ACIDS, WHICH ARE PRECIPITATED FROM SOLUTIONS THE PRECIPITATE *INSOLUBLE*

	HYDRIODIC ACID (HI). Iodide of Potassium.	**HYDROSULPHURIC ACID** (H_2S). Sulphide of Ammonium.
I. **Nitrate of Silver.**	$KI + AgNO_3$ = pale-yellow (crystalline) α.........AgI insol. in hydrate of ammonium β. insol. in dilute nitric acid γ.	$(NH_4)_2S + 2AgNO_3$ = black..............................Ag_2S insol. in hydrate of ammonium, and cold dilute nitric acid. sol. in boiling nitric acid, with separation of sulphur.
II. **Acetate of Lead.**	$KI + PbC_2H_3O_2$ = orange (crystalline)................PbI sol. in boiling water, and in chloride of ammonium. insol. in hydrate of ammonium.	$(NH_4)_2S + 2PbC_2H_3O_2$ = black α..............................Pb_2S insol. in cold water and dilute acids.
III. **Cuprous Sulphate.**	$2KI + (Cu_2)_2SO_4$ δ = white-brown..........................$2Cu_2I$ decomposed by nitric or sulphuric acid with evolution of free iodine.	[$(NH_4)_2S + 2Cu_2Cl$ β = $(Cu_2)_2S + 2NH_4Cl$] black
IV. **Cupric Sulphate.**	———	[$(NH_4)_2S + Cu_2SO_4$ = $Cu_2S + (NH_4)_2SO_4$] black
V. **Ferrous Sulphate.**	———	[$(NH_4)_2S + Fe_2SO_4$ = $Fe_2S + (NH_4)_2SO_4$] black
VI. **Ferric Chloride.**	———	[$3(NH_4)_2S + 2Fe_2Cl_3$ = $(Fe_2)_2S_3 + 6NH_4Cl$] black
VII. **Heated with Concentrated Sulphuric Acid.**	Solid iodide + H_2SO_4 **Evolution of iodine vapour.** (condenses, if much iodine, to black crystals.) $2KI + H_2SO_4 = 2I + K_2SO_4$* Test. Contact with starch paste on a glass rod; *the starch becomes blue* ζ.	Sulphide + H_2SO_4 or HCl γ. **Evolution of H_2S gas** δ. $(NH_4)_2S + H_2SO_4 = H_2S + (NH_4)_2SO_4$ Test. Contact with paper moistened with acetate of lead; *the paper is blackened.* (Pb_2S) *Odour.*
Heated with Concentrated Nitric Acid.	——— (See note ε.)	Solid sulphide + HNO_3 ε. **Separation of sulphur,** with formation of sulphuric acid. $2Fe_2S + 10HNO_3 = 2Fe_2(NO_3)_3$ ζ $+ 4H_2O + S + H_2SO_4 + 2N_2O_2$

α The ppt. becomes dark on exposure to light.
β Addition of hydrate of ammonium decolorises the solution.
γ Concᵈ. nitric acid decomposes the ppt. with separation of free iodine.
δ The cuprous salt is generally formed for this purpose by adding ferrous sulphate to cupric sulphate, thus, $2Cu_2SO_4 + 2Fe_2SO_4 = (Cu_2)_2SO_4 + (Fe_2)_2(SO_4)_3$.
ε Chlorine reacts with iodides as with bromides (see p. 50). The colour of the ether solution is violet or red-brown. The iodine may be converted into iodide of potassium, and tested by nitrate of silver.
Many other compounds decompose iodides, with evolution of free iodine, e.g. bromine, nitric acid, nitrous acid, hydrochloric acid, sulphuric acid, and peroxide of manganese. According as the iodine is in excess or less in quantity, so is it precipitated in the form of black crystals, or given off as violet vapours.
ζ If much iodine is present the colour is black.

* The intermediate reaction, with HI and H_2SO_4 occurs here, as in the case of HBr, Bromine (p. 50).

α This reaction is very characteristic; a piece of paper moistened with a soluble salt of lead (acetate) manifests it in presence of hydrosulphuric acid suspended in the gaseous form in the air.
β This salt may be formed by adding stannous chloride to a solution of cupric chloride, or by boiling metallic copper with an acid solution of cupric chloride.
γ Many sulphides are decomposed by *dilute* sulphuric or hydrochloric acid, e.g. sulphides of the alkalies, alkaline earths, iron, manganese and zinc. Others, as sulphides of cobalt, nickel, antimony, lead, require concentrated acid.
δ Some sulphides, e.g. copper, lead, do not evolve this gas when treated with sulphuric or hydrochloric acid.
ε Sulphide of mercury alone resists the decomposing action of concᵈ nitric acid. It only yields to the action of aqua-regia.
ζ Generally, the action of nitric acid upon decomposable sulphides, is to form the most highly oxidised combination of which the metal of the sulphide is susceptible; thus, sulphide of lead yields sulphate, sulphide of tin, binoxide.

SECTION II.

OF THEIR SALTS, BY *NITRATE OF SILVER.*
IN HYDRATE OF AMMONIUM.

HYDRO-FERRO-CYANIC ACID ($H_4FeCy_3 = H_4Cfy$). Ferrocyanide of Potassium.	HYDRO-SULPHO-CYANIC ACID ($HCNS = HCsy$)α. Sulphocyanide of Potassium.	
$K_4FeCy_3 + 2AgNO_3$ = white .... Ag_4FeCy_3 insol. in hydrate of ammonium. insol. in dilute nitric acid. decomposed by heat, with separation of metallic silver.	$KCNS + AgNO_3$ = white .... $AgCNS$ insol. in hydrate of ammonium. insol. in dilute nitric acid. decomposed by heat, with separation of metallic silver.	I. Argentic Salts.
$K_4FeCy_3 + 2PbC_2H_3O_2$ = white .... Pb_2FeCy_3 sol. in chloride of ammonium. insol. in water and hydrate of ammonium.	$KCNS + PbC_2H_3O_2$ = pale-yellow .... $PbCNS$ (separating slowly)β decomposed by water.	II. Plumbic Salts.
[$K_4FeCy_3 + 2Cu_2Cl$ = $(Cu_2)_2FeCy_3 + 2KCl$] white α	[$KCNS + Cu_2Cl$ = $Cu_2CNS + KCl$] white	III. Cuprous Salts.
$K_4FeCy_3 + Cu_2SO_4$ = red-brown .... Cu_2FeCy_3 insol. in acids, water, and salts of ammonium.	Forms with difficulty γ.	IV. Cupric Salts.
$4K_4FeCy_3 + 3Fe_2SO_4$ = white (rapidly becomes blue)..$2[KFe_2(FeCy_3)_3]$ nearly insol. in water. nitric acid converts it into *prussian blue*.	———	V. Ferrous Salts.
$3K_4FeCy_3 + 2Fe_2Cl_6$ = blue = prussian blue....$(Fe_2)_2(FeCy_3)_3$ insol. in water. sol. in nitric acid.	Soluble, blood-red colour of solution characteristic. (see Hydrocyanic Acid, p. 51.)	VI. Ferric Salts.
Solid ferrocyanide β + H_2SO_4 Evolution of carbonic oxide γ. $\left.\begin{matrix} 2K_4FeCy_3 + 6H_2O \\ + \\ 6H_2SO_4 \end{matrix}\right\} = \left\{\begin{matrix} 6CO + 3(NH_4)_2SO_4 \\ + \\ 2K_2SO_4 + Fe_2SO_4 \end{matrix}\right.$ Test. *Blue flame when ignited.* ————	Solid sulphocyanide + H_2SO_4. Evolution of carbonic oxide, with separation of sulphur. $\left.\begin{matrix} 2KCNS + 2H_2O \\ + \\ 2H_2SO_4 \end{matrix}\right\} = \left\{\begin{matrix} 2CO + 2S \\ + \\ K_2SO_4 + (NH_4)_2SO_4 \end{matrix}\right.$ ———— Yellow ppt. (composition of the ppt. uncertain δ.)	VII. Decompositions.

α The ppt. rapidly becomes brown, being converted into a cupric salt.
β Ferrocyanide of sodium also reacts in this manner.
γ With *dilute* sulphuric acid, hydrocyanic acid is evolved from alkaline ferrocyanides, recognised by its odour. The decompositions are not to be relied upon for detecting this acid radical. The reactions of its solutions with ferrous and ferric salts are chiefly employed to distinguish it.

Northcote and Church give the following interesting decomposition:—Boil a ferrocyanide (not alkaline) with hydrates or carbonates of potassium or sodium—ferrocyanide of potassium or sodium is thus formed—filter—evaporate the solution to dryness—ignite—a fused mass is obtained—in that portion of the mass soluble in water an *alkaline cyanide* will be found if tested for (see Hydrocyanic Acid, p. 51); in that portion insoluble in water, after washing and dissolving in hot hydrochloric or nitric acid, *iron* may be tested for*.

* *Manual of Qualitative Chemical Analysis*, p. 345.

α This acid is found in the saliva.
β Converted by hot nitric acid into plumbic sulphate.
γ It may be produced in a concentrated solution of sulphocyanide of potassium, by adding sulphuric acid, and then a saturated solution of cupric sulphate. The ppt. is black, the formula is CuCNS.
δ Chlorine gas passed into the solution of a sulphocyanide produces a similar ppt.

GROUP III.

ACIDS, WHICH ARE NOT PRECIPITATED FROM SOLUTIONS OF THEIR SALTS, BY *CHLORIDE OF BARIUM* OR *NITRATE OF SILVER*.

	NITRIC ACID (HNO_3). Nitrate of Potassium.	**CHLORIC ACID** ($HClO_3$)α. Chlorate of Potassium.	
I. **Ferrous Sulphate.**	$\left.\begin{array}{c}2KNO_3\\+\\4H_2SO_4\end{array}\right\} + 10Fe_2SO_4\beta =$ brown compound............$4Fe_2SO_4.N_2O_2$	———	I.
II. **Sulpindi-gotic Acid.**	$KNO_3 + H_2SO_4$ + sulpindigotic acid γ. Colour becomes yellow.	$KClO_3 + H_2SO_4$ + sulpindigotic acid. Colour becomes yellow.	II.
III. **Conc[d] Sulphuric Acid.**	i. Solid nitrate + H_2SO_4 + metallic copper. Evolution of red fumes δ. $\left.\begin{array}{l}2KNO_3+6Cu\\+4H_2SO_4\end{array}\right\}=\left\{\begin{array}{l}K_2SO_4+3Cu_2SO_4+4H_2O+\\N_2O_2\text{ (nitric oxide)}\end{array}\right.$ and $N_2O_2 + O_2 = N_2O_4$ (peroxide).	i. Solid chlorate + H_2SO_4 (not heated). Evolution of green-yellow gas β. (hypochloric anhydride.) $\left.\begin{array}{l}3KClO_3\\+2H_2SO_4\end{array}\right\}=\left\{\begin{array}{l}Cl_2O_4+KClO_4+\\2KHSO_4+H_2O\,\gamma.\end{array}\right.$	III. Decompo-sitions.
Hydrochlo-ric Acid.	ii. Solid nitrate + HCl. Evolution of red fumes and chlorine ε. $2KNO_3 + 4HCl = N_2O_4 + Cl_2 + 2KCl + 2H_2O$ Test. *Bleaching power of chlorine.*	ii. Solid chlorate + HCl (not heated). Evolution of yellow gas. (euchlorine δ.) $4KClO_3 + 12HCl = (9Cl + 3ClO_2) + 4KCl + 6H_2O$ Tests. *Odour. Explodes in contact with flame.*	
IV. **Sulphurous Acid.**	———	iii. $KClO_3$ + sulphurous acid. Chlorate is decomposed. Test. Add nitrate of silver; *white ppt.* ε, *insoluble in nitric acid* ζ.	IV.

α All nitrates are soluble in water, therefore they are not precipitated by reagents.

β If a crystal of ferrous sulphate is dropped into the solution of nitrate and acid, a brown ring forms round the crystal. Or the crystal may be dissolved (without heat) in the solution of nitrate, and the acid carefully added, when a brown stratum will appear at the junction of the two liquids.

γ A few drops only of solution of indigo should be used.

δ Nitric oxide is first formed, but becomes oxidised by exposure to air, is converted into peroxide, and appears in the form of red fumes.

ε The solvent powers of nitro-hydrochloric acid (aqua regia) depend on the ready decomposition of nitric acid by hydrochloric acid, and simultaneous evolution of free chlorine.

$$2HNO_3 + 2HCl = N_2O_4 + 2H_2O + 2Cl.$$

The nascent chlorine exerts a powerfully solvent action, greater than that of nitric or hydrochloric acid separately.

α All chlorates are soluble in water, hence they cannot be identified by the formation of precipitates.

β The application of heat, in performing this experiment, should be avoided, and small quantities only of chlorate used, since the explosion which attends the decomposition at a high temperature is liable to be dangerous; without heat, a slight crackling occurs during the decomposition, the solution at the same time becoming intensely yellow.

Chlorates do not exhibit the reaction with metallic copper which is observed in the case of nitrates.

γ This is the chief insoluble salt of the perchloric radical. Perchlorates are sufficiently distinguished from chlorates by decomposing with difficulty and not suffering decomposition by sulphurous acid. (See above IV. and note ζ.)

δ This gas explodes violently when heated.

ε The ppt. is a mixed sulphite and chloride of silver.

ζ The ppt. produced by nitrate of silver in a solution of a *perchlorate* treated with sulphurous acid consists entirely of sulphite of silver, and is soluble in nitric acid.

ORGANIC ACIDS.

GROUP I.

ACIDS, WHICH ARE PRECIPITATED FROM SOLUTIONS OF THEIR SALTS, BY *CHLORIDE OF CALCIUM.*

	TARTARIC ACID ($H_2C_4H_4O_6$). Tartrate of Sodium.	**CITRIC ACID** ($H_3C_6H_5O_7$). Citrate of Sodium.	
I. **Chloride of Calcium.**	$Na_2C_4H_4O_6 + 2CaCl$ = white (crystalline) α............$Ca_2C_4H_4O_6$ sol. in hydrate of potassium, reppt. on boiling β. sol. in acetic acid.	$Na_3C_6H_5O_7 + 3CaCl$ = white (crystalline) α............$Ca_3C_6H_5O_7$ insol. in hydrate of potassium. sol. in acetic acid.	I. Calcium Salts.
II. **Hydrate of Calcium.**	$Na_2C_4H_4O_6 + 2CaHO$ = white (flocculent, then crystalline)..$Ca_2C_4H_4O_6$ sol. in tartaric acid and chloride of ammonium, separating again after some time in the crystalline form.	No ppt. in *cold* solutions; a ppt. forms in *boiling* solutions, dissolving when cold.	II. Calcium Salts.
III. **Chloride of Barium.**	$Na_2C_4H_4O_6 + 2BaCl$ = white............................$Ba_2C_4H_4O_6$ sol. in salts of ammonium (except the hydrate). sol. in hydrochloric acid.	$Na_3C_6H_5O_7 + 3BaCl$ = white............................$Ba_3C_6H_5O_7$ sol. in salts of ammonium. sol. in acids and in much water.	III. Barium Salts.
IV. **Nitrate of Silver.**	$Na_2C_4H_4O_6 + 2AgNO_3$ = white (crystalline)............$Ag_2C_4H_4O_6$ sol. in hydrate of ammonium γ and in acids. insol. in water.	$Na_3C_6H_5O_7 + 3AgNO_3$ = white (flocculent)..............$Ag_3C_6H_5O_7$ sol. in hydrate of ammonium β and in nitric acid. sol. in boiling water.	IV. Argentic Salts.
V. **The solid salt, heated with Conc^d H_2SO_4.**	**Blackening** (separation of carbon) and **evolution of carbonic oxide δ.**	**Blackening** *only on prolonged boiling,* and **evolution of carbonic oxide γ.**	V. Decompositions.

α Presence of salts of ammonium prevents precipitation; free ammonia *promotes* it. If hydrate of ammonium is added to a small quantity of tartrate of calcium, and a small crystal of nitrate of silver then added, on cautiously heating the sides of the test-tube become coated with metallic silver.

β Tartrate of calcium, after being re-precipitated from the boiling solution in hydrate of potassium, again re-dissolves on cooling.

γ On boiling, this solution becomes black, from the reduction of silver to the metallic state.

δ Sulphurous and carbonic anhydrides are simultaneously evolved. An odour of burnt sugar is apparent on heating.

α In dilute solutions, the ppt. only appears on boiling, citrate of calcium being less soluble in hot than in cold water.

No ppt. is produced with citric acid itself; this must therefore be neutralised by salts of potassium or sodium.

Presence of salts of ammonium prevents precipitation; free ammonia *promotes* it.

β No blackening occurs on boiling.

γ Sulphurous and carbonic anhydrides are also evolved. Acetic acid is also one of the products of decomposition.

ORGANIC

GROUP

ACIDS, WHICH ARE PRECIPITATED FROM *NEUTRAL*

	SUCCINIC ACID ($H_2C_4H_4O_4$). Succinate of Ammonium.	**BENZOIC ACID** ($HC_7H_5O_2$). Benzoate of Ammonium.
I. **Ferric Chloride.**	$3(NH_4)_2C_4H_4O_4 + 2Fe_2Cl_6 =$ red-brown α .... $(Fe_2)_2(C_4H_4O_4)_3$? sol. in hot acetic or succinic acid β and in mineral acids. decomposed by hydrate of ammonium.	$3(NH_4.C_7H_5O_2) + Fe_2Cl_6 =$ buff α .... $Fe_2(C_7H_5O_2)_6$? sol. in most acids. decomposed by hydrate of ammonium.
II. **Nitrate of Silver.**	$(NH_4)_2C_4H_4O_4 + 2AgNO_3 =$ white .... $Ag_2C_4H_4O_4$ sol. in hydrate of ammonium and nitric acid. sol. sparingly in water.	$NH_4.C_7H_5O_2 + AgNO_3 =$ white .... $AgC_7H_5O_2$ sol. in hydrate of ammonium and nitric acid. sol. in boiling water.
III. **Acetate of Lead.**	$(NH_4)_2C_4H_4O_4 + 2PbC_2H_3O_2 =$ white .... $Pb_2C_4H_4O_4$ sol. in nitric acid. insol. in alcohol.	$NH_4.C_7H_5O_2 + PbC_2H_3O_2 =$ white (after some time) β.... $PbC_7H_5O_2$
IV. **Chloride of Barium.**	$(NH_4)_2C_4H_4O_4 + 2BaCl =$ white .... $Ba_2C_4H_4O_4$ *if alcohol is previously added to the solution* γ.	soluble, even in alcohol.
V. **The solid salt heated with Concd H_2SO_4.**	**Volatile,** without decomposition. **No blackening.**	**Volatile,** without decomposition. **No blackening.**
Solution heated with KHO.	———	———
VI. **Solution of Gelatine.**	———	———
	α The composition of this ppt. is variable. β Hence in a solution containing an excess of acid, ferric chloride produces no ppt. γ If the solution to be tested consists of free succinic acid it must be neutralised by hydrate of ammonium before adding alcohol and chloride of barium.	α The whole of the iron is precipitated if ammonium is present as above. Hence benzoate of ammonium is employed to separate ferrous and ferric salts from a solution in which both are present. The iron may then be removed from the precipitate by hydrochloric acid, leaving benzoic acid undissolved. Or the latter may be removed by hydrate of ammonium, leaving the iron as ferric hydrate. β If benzoic acid is combined with potassium or sodium instead of ammonium, a white flocculent ppt. is thrown down. The ppt. formed above is very slight even after the lapse of some time.

ACIDS.

II.

SOLUTIONS OF THEIR SALTS, BY *FERRIC CHLORIDE.*

GALLIC ACID ($H_3C_7H_3O_5$). Gallic Acid.	**TANNIC ACID** ($H_3C_{27}H_{19}O_{17}$). Tannic Acid.	
$H_3C_7H_3O_5 + Fe_2Cl_6$ = bluish-black solution α .... $Fe_2C_7H_3O_5$	$H_3C_{27}H_{19}O + Fe_2Cl_6$ = bluish-black α .... $Fe_2C_{27}H_{19}O_{17}$	I. Ferric Salts.
————	$H_3C_{27}H_{19}O_{17} + 3AgNO_3$ = red-brown .... $Ag_3C_{27}H_{19}O_{17}$	II. Argentic Salts.
$H_3C_7H_3O_5 + 2PbC_2H_3O_2$ = white β .... $Pb_2HC_7H_3O_5$ sol. in conc[d]. acetic acid.	$H_3C_{27}H_{19}O_{17} + 3PbC_2H_3O_2$ = white .... $Pb_3C_{27}H_{19}O_{17}$?	III. Plumbic Salts.
$H_3C_7H_3O_5 + BaCl$ = white .... $BaH_2C_7H_3O_5$ sol. sparingly in water.	$H_3C_{27}H_{19}O_{17} + 3BaCl$ = white .... $Ba_3C_{27}H_{19}O_{17}$ insoluble in water.	IV. Barium Salts.
Blackening. (separation of carbon.)	Dark purple liquid.	V. Decompositions.
Brown coloration. *immediately* γ.	Brown coloration β.	
————	White viscous ppt. γ.	VI. Precipitation of Gelatine.

α On boiling, the liquid decolorises with formation of a ferrous salt and escape of carbonic anhydride.
β This ppt. also contains some water of crystallisation. If the acetate of lead is in excess the gallic acid is completely precipitated.
γ From rapid oxidation and formation of a brown colouring matter, the colour first appears yellow.

α This constitutes the colouring matter of ink.
β Oxidation is not so rapid as in the case of gallic acid, hence the coloration is not so immediate.
γ This reaction is characteristic of tannic acid. Upon it is based the use of bark (which contains tannic acid) in tanning animal skins.

ORGANIC ACIDS.

GROUP III.

ACIDS, WHICH ARE *NOT* PRECIPITATED FROM THEIR SOLUTIONS, BY *CHLORIDE OF CALCIUM OR FERRIC CHLORIDE.*

	ACETIC ACID ($HC_2H_3O_2$). Acetate of Potassium.	**URIC ACID** ($H_2C_5N_4H_2O_3$). Uric Acid.	
I. **Nitrate of Silver.**	$KC_2H_3O_2 + AgNO_3$ = white (crystalline)............$AgC_2H_3O_2$ sol. in hydrate of ammonium. sol. in boiling water, reppt. on cooling.	White ppt. (becomes *black*, if heated.)	I. Argentic Salts.
II. **Ferric Chloride.**	$3KC_2H_3O_2 + Fe_2Cl_6$ = red-brown solution α.....$Fe_2(C_2H_3O_2)_3$ on boiling, the iron is *precipitated* as basic ferric acetate β.	———	II. Ferric Salt.
III. **The solid salt heated with Conc^d^ H_2SO_4.**	H_2SO_4 + acetate + alcohol. Aromatic odour γ.	Brown Solution. (Water reprecipitates uric acid.)	III. Decomposition.
IV. **Hydrate of Ammonium.**	———	Solid uric acid dissolved in nitric acid; solution evaporated to dryness; hydrate of ammonium added in excess. Purple colour α.	IV. Murexide.

Acetic acid notes:

α Addition of hydrochloric acid to a solution which appears red from presence of ferric acetate, renders it yellow. This reaction serves to distinguish between ferric acetate and ferric sulphocyanide (see p. 53).

β On boiling, the solution becomes colourless.

γ The odour is due to the formation of acetic ether. Thus

$$KC_2H_3O_2 + \underset{\text{alcohol}}{C_2H_5HO} = \underset{\text{acetic ether.}}{C_2H_5C_2H_3O_2} + KHO$$

Uric acid notes:

α This coloration is due to the formation of a body termed "Murexide."

The uric acid is dissolved by the aid of heat, the solution being attended with effervescence and evolution of red fumes. This arises from oxidation of uric acid at the expense of nitric acid.

PART III.

SYSTEMATIC COURSE OF ANALYSIS

FOR

BASIC RADICALS.

ANALYSIS OF A SALT OR MIXTURE OF SALTS FOR BASIC RADICALS.

The Course of Analysis is divided into three stages, or series of processes.

I. The preliminary examination. (The substance under analysis being in the solid form.)

The general plan of procedure is detailed on pages 62 and 63. The experiments should be performed in the order in which they are placed. The beginner is cautioned against allowing the results of his experiments to determine him in selecting subsequent experiments, e.g. If he fancies he has obtained the dark-green residue, due to cobalt, as in Exp. II. p. 62, he should not *at once* proceed to seek for more confirmatory evidence of the presence of cobalt by fusing the substance with borax on platinum wire.

By carefully performing each of the five experiments (the sixth is more special) in order, every time a preliminary examination is undertaken, the student will obtain what is of greater moment to him than the mere detection of a radical, accuracy in observing and facility in manipulating.

The results of this examination are to be carefully noted, but in no case must the results be allowed to influence the subsequent method of analysis in the wet way. Inferences from an inspection of the physical properties of a substance should also be carefully *avoided* by the beginner.

II. The process of solution: see pages 20 and 21.

It is useful for the purposes of analysis to classify substances according to their deportment with solvents, thus—

1st Class—soluble in water.

2nd ... insoluble in water, but soluble in acids.

3rd ... insoluble in water or acids.

The methods of analysis as detailed in the following pages are framed for the first class, but such modifications as may be required in the case of bodies which belong to the second class, are readily introduced. A separate scheme for the analysis of substances insoluble in water or acids is appended to the Fourth Part.

In endeavouring to effect the solution of a body which is insoluble in water, in *hydrochloric acid*, certain reactions may occur which the analyst should carefully observe. (See p. 64.)

[If *organic matter* is present in the substance under analysis (recognized by the charring in the preliminary examination) it should be removed, since otherwise it would interfere with the detection of certain basic radicals. This may generally be effected, if no volatile metal is present, by igniting the substance, to carbonize the organic matter. If the residue is soluble it may be removed from the carbonaceous mass by filtration. If a volatile metal is present, organic matter may be destroyed by means of such oxidising agents as nitric acid and chlorate of potassium.]

In cases where only partial solution is effected by a solvent, the solution should be separated from the residue and examined in order to ascertain the nature of that which has been dissolved.

III. The application of reagents to the solution.

As before stated (Introduction) reagents are *general* or *special*, according to the purposes they serve in the course of analysis.

The results of the application of *general reagents* to a solution are tabulated on pages 64 and 65. The Table furnishes the analyst with a comprehensive view of reactions which may occur under different circumstances, in using these reagents, and indicates the contingencies for which he must provide.

In the Methods of Analysis for the several groups (which apply to the detection of more than one basic radical in a solution) almost every contingency which would be likely to arise in the course of analysis has been considered; hence, some complexity has resulted. A summary has been introduced on pages 66 and 67, by means of which the several steps may be traced. The student should thoroughly apprehend the rationale of each process, as detailed in the centre of each page, before attempting to perform the manipulations in the right-hand marginal columns.

The following are rules of essential importance in reference to the analytical course.

Economy of time should be aimed at. During the evaporation of a filtrate, a precipitate may be examined. Many operations may be carried on simultaneously. On the other hand, haste and attempts to invent short roads will invariably preclude success.

Economy in the use of material is essential. This is especially the case where only a limited quantity of the substance to be analyzed is at the disposal of the operator. Reagents should invariably be used sparingly, unless the contrary is distinctly enjoined, and applied cautiously, even by drops, otherwise valuable results may be obscured.

Precipitation should be complete. To ensure this, small quantities of any reagent should be added *gradually*, and the precipitate allowed to subside between each addition; this should be repeated until no further precipitate is produced in the clear supernatant liquid.

The washing of precipitates should be thoroughly performed. Beginners often neglect this precaution, and hence they involve themselves in endless complexities. It is evidently of fundamental importance that a precipitate containing members of one group should be completely freed from a solution which may contain members of other groups.

The greatest care should be exercised to perform *thoroughly* all the operations of analysis, such as boiling, evaporation, &c. It will be a good plan to test specially in order to ascertain if the desired result has been accomplished, in a given case, e.g. in boiling off hydrosulphuric acid from a solution, a piece of paper moistened with acetate of lead and held over the test tube will afford the necessary information.

The student is urged to determine what he proposes to effect, before undertaking any experiments; in the event of failure, to thoroughly investigate at once the possible causes of non-success, and to repeat experiments with any modifications experience may suggest, again and again, until success is achieved.

PRELIMINARY EXAMINATION.

(BASIC RADICALS.)

Exp. I.		
i. A portion of the powdered substance is heated in a **Glass tube open at both ends** and held obliquely in the flame	Aqueous vapour is expelled, condensing on a cool part of the tube.	
	Alkaline.	NH_4.
	Acid.	A volatile acid.
	Fumes or gases are expelled.	
	The gas kindles burning wood.	O. (chlorate or nitrate).
	has the odour of S. (acid).	SO_2 (sulphate).
	burns with a blue flame.	CO. (oxalate).
	has the odour of Cy, and burns with a crimson flame.	Cy. (cyanide).
	has the odour of H_2S.	H_2S. (sulphide).
	A sublimate forms on a cool part of the tube.	
	Metallic lustre.	As. Hg. Cd.
	Whitish.	Sb. Bi.
ii. A portion of the powdered substance is *mixed with Na_2CO_3* and heated in a **Glass tube closed at one end.**	A sublimate forms on a cool part of the tube.	
	Grey globules.	Hg.
	A black mirror.	As (see p. 14).
Exp. II.		
A portion of the powdered substance is heated on **Charcoal.**	Volatile. (Incrustation	
	White.	Hg. As (garlic odour). NH_4.
	Red-brown.)	Cd. [Confirm NH_4 by KHO.
i. In the *outer* blowpipe flame. (Coloration of flame, as in Exp. V. is also observed here.)	Non-volatile. Coloured incrustation.	
	Orange (hot). Yellow (cold).	Pb.
	Bluish-white and distant.	Sn. Sb. Sn_2 (dense white).
	Infusible *coloured* residue.	
	Brown mass (yellow when cold).	Bi.
	Green.	Cr.
	Brown (hot). Red (cold).	Fe.
	Brown.	Mn.
	Yellow (hot). White (cold).	Zn.
	Dark-green.	Co.
	Greenish or Dull black.	Ni.
	Infusible *incandescent* residue.	Ca. Mg. Zn. Al.
ii. In the *inner* blowpipe flame.	Coloured incrustation.	
	Yellow.	Pb.
	Red-brown.	Cd.
	White or Yellowish-white.	Zn.
iii. Fused *with Na_2CO_3* in the *inner* blowpipe flame.	Reduction of metal.	
	Metallic globules are white, malleable.	Ag. Pb. Sn.
	coloured.	Au (yellow). Cu (reddish).
	coloured, brittle.	Sb. Bi.
	Black metallic mass.	Pt.

PRELIMINARY EXAMINATION (CONTINUED).

(BASIC RADICALS.)

Exp. III. The residue from Exp. II. i. is moistened with **Nitrate of Cobalt,** and again heated.	The mass becomes coloured.	Bluish-green. Blue. Green. Pale-pink.	Sn. Al. Zn. Mg.
Exp. IV. A portion of the powdered substance is fused with **Borax, on Platinum wire.** (a very small quantity of the powdered substance should be used in this Exp.)	A bead is formed. *In Outer flame.* Yellow (hot) } Colourless (cold) } Transparent and clear. Green (hot) } Blue (cold) } Green. Red (hot) } Yellow or Colourless (cold) } Amethyst. Clear. Blue. Violet (hot) } Red-brown (cold). }	*In Inner flame.* Colourless (hot) } Brick-red (cold) } Green. Yellow (hot) } Bottle-green (cold) } Colourless. Milk-white. Blue. Grey.	Bi. Cd. Cu. Cr. Fe. Mn. Zn. Co. Ni.
Exp. V. A portion of the powdered substance is heated *alone*, on **Platinum wire or foil,** in the outer blowpipe flame. (also see Exp. II. i.)	The flame is coloured.	Yellowish-green. Crimson. Orange-red. Violet. Yellow (intense). Carmine.	Ba. Sr. Ca. K. Na. (characteristic). Li.
Exp. VI. A portion of the powdered substance *mixed with* $Na_2CO_3 + KCy$ is fused on Platinum foil in the *inner* blowpipe flame.	The mass is coloured.	Bluish-green (cold).	Mn. (characteristic).

The presence of *organic* **matter** in a substance is usually revealed by the charring and blackening (from separation of carbon) which takes place when the substance is subjected to a high temperature.

BEHAVIOUR OF BASIC RADICALS AND CERTAIN ACID RADICALS WITH THOSE

Basic Radicals and salts.		Hydrochloric Acid in an acid, alkaline or neutral solution.	Hydrosulphuric Acid in an acid solution.
I.			
Silver	1.	1. **Chloride,** white . . ac. neut. or alk. α sol.	
Lead	2.	2. **Chloride,** white . . ac. neut. or alk. α sol. 2. Sulphate . . . . alk sol.	2. *Sulphide* in dilute, not too acid solutions.
Mercury (Hg_2)	3.	3. **Chloride,** white . . ac. alk. or neut. sol.	3. Sulphide.
II.			**All as Sulphides.**
Tin (Sn_2)	1.		1. Dark brown.
,, (Sn)	2.	2. Sulphide, with or without evolution of gas γ alk. sol.	2. Yellow.
Antimony (Sb_2)	3.	3. Sulphide γ . . . . alk. sol.	3. Orange.
,, (Sb)	4.	4. Sulphide, with or without evolution of gas γ alk. sol. 4. *Oxychloride β* . . . . . . *ac. or neut. sol.*	4. Orange-yellow.
Arsenic (As_2)	5.		5. Orange.
,, (As)	6.	6. Sulphide, with or without evolution of gas γ alk. sol.	6. Yellow.
Platinum	7.	7. Sulphide, with evolution of gas γ . . . . alk. sol.	7. Brown-black.
Gold	8.	8. Sulphide ,, ,, γ . . . . alk. sol.	8. Black.
Mercury (**Hg**)	9.	9. Sulphide ,, ,, γ . . . . alk. sol.	9. White, finally black.
Bismuth	10.	10. *Oxychloride β* . . . . . *ac. or neut. sol.*	10. Brown black.
Cadmium	11.		11. Brilliant yellow.
Copper	12.	12. Sulphide, with evolution of gas γ . . . . alk. sol.	12. Black.
III.			
Aluminium	1.		
Chromium	2.		2 *Reduced to Sesquioxide. S separates α.*
Iron (Fe_2)	3.		3. *Reduced to Ferrous salt. S separates β.*
,, (Fe)	4.		
IV			
Manganese	1.		
Zinc	2.		
Cobalt	3.		
Nickel	4.	4. Sulphide, with evolution of gas γ . . . alk. sol.	
V.			
Barium	1.	1. Chloride . . . . saturated ac. or neut. sol.	1. Sulphate} From oxidation of Sulphur, *when*
Strontium	2.		2. Sulphate} *Nitric acid is present.*
Calcium	3.		
Magnesium	4.		
VI.			
Potassium	1.		
Sodium	2.	Not precipitated.	Not precipitated.
Ammonium	3.		
Lithium	4.		
Acid Radicals and salts.			
Carbonic	1.	1. *Evolution of gas γ δ* . . . . . alk. sol.	
Silicic	2.	2. White or yellow ppt. α . . . . . alk. sol.	
Boracic	3.	3. Precipitated from a *conc^d* solution. Soluble in water	
Sulphurous	4.		4. Separation of Sulphur.
Hydrochloric	5.		5. ,, ,, (if Chlorine is free).
Hydrobromic	6.		6. ,, ,, (if Bromine is free).
Iodic	7.		7. ,, ,, (reduced to Iodide).
Hydrocyanic	8.	8. *Evolution of gas γ δ* . . . . . alk. sol.	
Hydrosulphuric	9.	9. *Evolution of gas γ δ* . . . . . alk. sol.	
Nitric	10.		10. Separation of Sulphur (from oxidation).

α Treat as a substance insoluble in water and acids, as in Appendix.
β These re-dissolve in an excess of hydrochloric acid.
γ See for all these cases, p. 69, where directions are given for dealing with them according to Fresenius.
δ The presence of these acids is generally revealed by the effervescence which attends their decomposition by hydrochloric acid.

α Recognised by greenish colour of the solution.
β Yellowish-white ppt. of sulphur.

REAGENTS EMPLOYED TO SEPARATE THE BASIC RADICALS INTO GROUPS.

Hydrate of Ammonium in presence of NH_4Cl and HNO_3.	Sulphide of Ammonium.	Carbonate of Ammonium in presence of NH_4HO and NH_4Cl.
1. **Hydrate,** white. 2. **Hydrate,** bluish-green. 3. **Hydrate,** white yellow. 4. a. As *Phosphates* also, in presence of the phosphoric radical.	3. Sulphide.	
1. Hydrate (only partially).	1. **Sulphide,** flesh colour. 2. **Sulphide,** white. 3. **Sulphide,** black. 4. **Sulphide,** black.	
1., 2., 3. as *Phosphates, Oxalates, Borates* or *Fluorides,* in presence of the corresponding acid radicals. 4. *Phosphate* in presence of the phosphoric radical.		1. **Carbonate,** white. 2. **Carbonate,** white. 3. **Carbonate,** white. 4. a.
Not precipitated.	Not precipitated.	Not precipitated.
a Any ferrous salt has been converted into a ferric salt by the addition of nitric acid, and boiling.		a Magnesium is not precipitated by carbonate of ammonium in presence of chloride of ammonium.

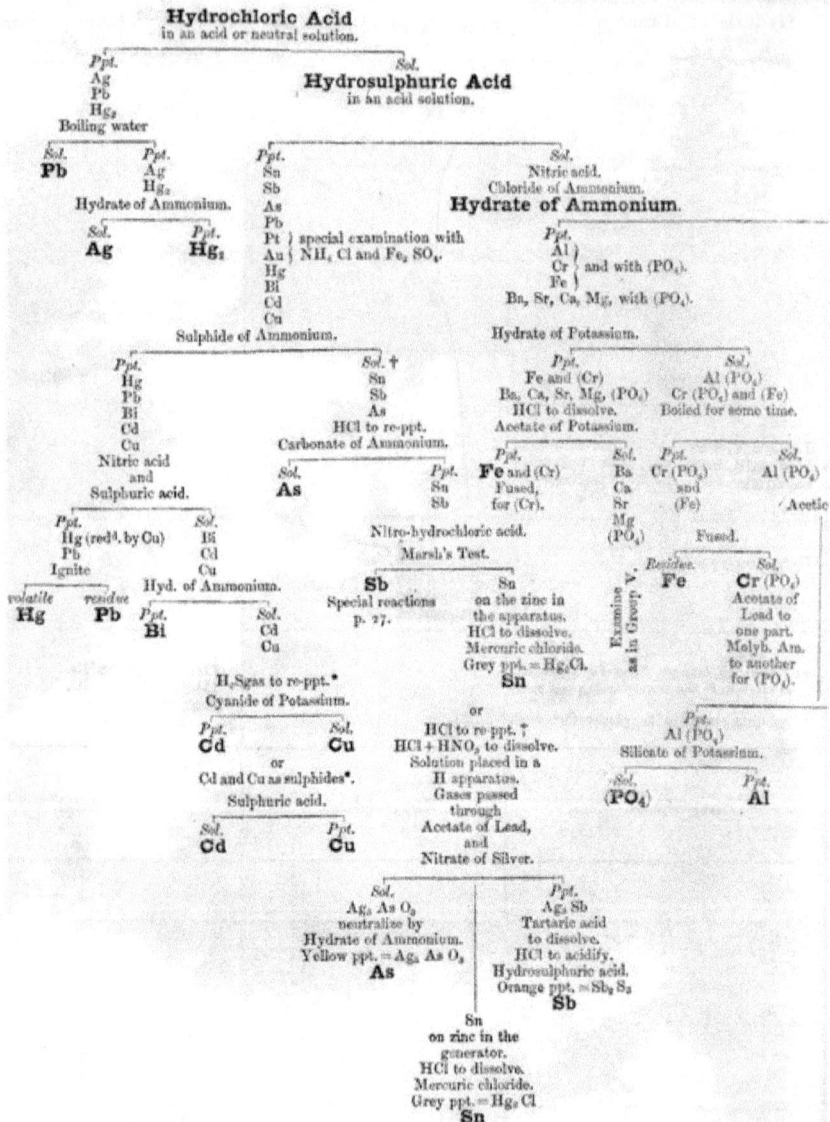

Hydrochloric Acid
in an acid or neutral solution.

- *Ppt.* Ag, Pb, Hg_2 — Boiling water
  - *Sol.* **Pb**
  - *Ppt.* Ag, Hg_2 — Hydrate of Ammonium.
    - *Sol.* **Ag**
    - *Ppt.* **Hg_2**
- *Sol.*

Hydrosulphuric Acid
in an acid solution.

- *Ppt.* Sn, Sb, As, Pb, Pt } special examination with Au } NH_4 Cl and Fe_2 SO_4. Hg, Bi, Cd, Cu — Sulphide of Ammonium.
  - *Ppt.* Hg, Pb, Bi, Cd, Cu — Nitric acid and Sulphuric acid.
    - *Ppt.* Hg (red^d. by Cu), Pb — Ignite
      - *volatile* **Hg**
      - *residue* **Pb**
    - *Sol.* Bi, Cd, Cu — Hyd. of Ammonium.
      - *Ppt.* **Bi**
      - *Sol.* Cd, Cu — H_2S gas to re-ppt.* Cyanide of Potassium.
        - *Ppt.* **Cd**
        - *Sol.* **Cu**
        - or Cd and Cu as sulphides*. Sulphuric acid.
          - *Sol.* **Cd**
          - *Ppt.* **Cu**
  - *Sol.* † Sn, Sb, As — HCl to re-ppt. Carbonate of Ammonium.
    - *Sol.* **As**
    - *Ppt.* Sn, Sb — Nitro-hydrochloric acid. Marsh's Test.
      - **Sb** Special reactions p. 27.
      - Sn on the zinc in the apparatus. HCl to dissolve. Mercuric chloride. Grey ppt. = Hg_2Cl. **Sn**
      - or HCl to re-ppt. † HCl + HNO_3 to dissolve. Solution placed in a H apparatus. Gases passed through Acetate of Lead, and Nitrate of Silver.
        - *Sol.* Ag_3 As O_3 neutralize by Hydrate of Ammonium. Yellow ppt. = Ag_3 As O_3 **As**
        - *Ppt.* Ag_3 Sb Tartaric acid to dissolve. HCl to acidify. Hydrosulphuric acid. Orange ppt. = Sb_2 S_3 **Sb**
        - Sn on zinc in the generator. HCl to dissolve. Mercuric chloride. Grey ppt. = Hg_2 Cl **Sn**
- *Sol.* Nitric acid. Chloride of Ammonium.

Hydrate of Ammonium.

- *Ppt.* Al, Cr, Fe } and with (PO_4). Ba, Sr, Ca, Mg, with (PO_4). — Hydrate of Potassium.
  - *Ppt.* Fe and (Cr), Ba, Ca, Sr, Mg, (PO_4) — HCl to dissolve. Acetate of Potassium.
    - *Ppt.* **Fe** and (Cr) Fused, for (Cr).
    - *Sol.* Ba, Ca, Sr, Mg, (PO_4) — Examine as in Group V.
  - *Sol.* Al (PO_4), Cr (PO_4) and (Fe) — Boiled for some time.
    - *Ppt.* Cr (PO_4) and (Fe) — Fused.
      - *Residue.* **Fe**
      - *Sol.* **Cr** (PO_4) Acetate of Lead to one part. Molyb. Am. to another for (PO_4).
    - *Sol.* Al (PO_4) — Acetic
      - *Ppt.* Al (PO_4) — Silicate of Potassium.
        - *Sol.* **(PO_4)**
        - *Ppt.* **Al**

URSE OF ANALYSIS FOR BASIC RADICALS.

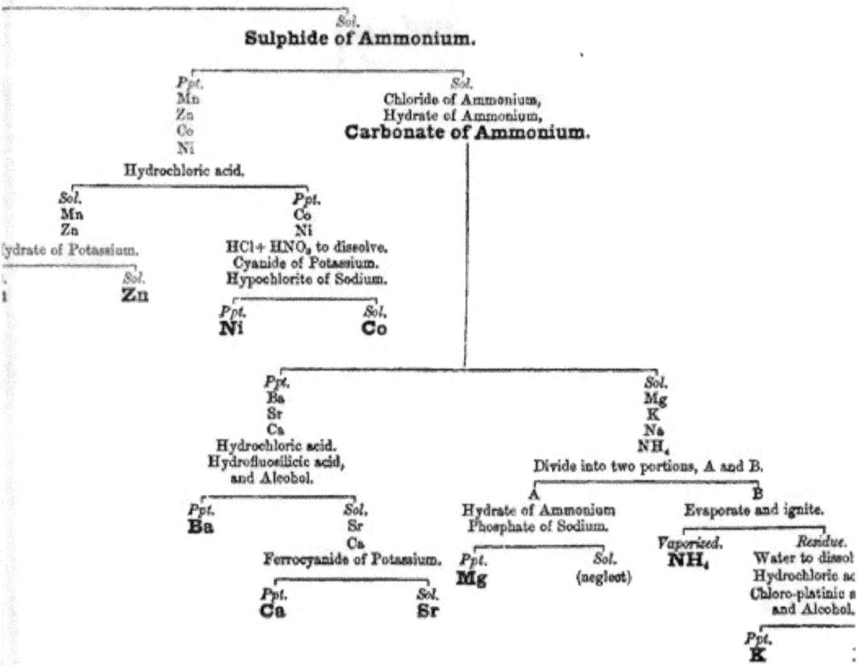

METHOD OF ANALYSIS FOR GROUP I.

HYDROCHLORIC ACID

precipitates from *acid* or *neutral* solutions,

Chlorides of {Lead, Silver.
Mercurous Chloride.

LEAD is not precipitated by hydrochloric acid, from *dilute* solutions, because chloride of lead is soluble in much water. If present it will be detected during the course of the analysis for the next group.

OXYCHLORIDES OF ANTIMONY AND BISMUTH may be precipitated along with members of this group. They re-dissolve in an excess of hydrochloric acid.

EFFERVESCENCE AND EVOLUTION OF GAS, on addition of hydrochloric acid, reveals the presence of the carbonic, hydrosulphuric, or hydrocyanic, acid-radicals. They will be readily recognised by their characteristic reactions.

Hydrochloric acid must be added drop by drop. A few drops will determine the presence or absence of members of this group; if present, excess must be added; if absent, the only object to be attained is the acidification of the solution preparatory to treating it with hydrosulphuric acid.

ANALYSIS OF PRECIPITATE PRODUCED BY HYDROCHLORIC ACID, IN AN ACID OR NEUTRAL SOLUTION.

	The *filtrate*, which may contain members of other groups, is set aside for further examination. The *precipitate* is thoroughly washed with *cold water*. If the washings *are added to the filtrate*, any appearance of turbidity will reveal the presence of Antimony or Bismuth.	*The solution is acid or neutral*, Add HCl in excess. 1 ....Filter. (Examine filtrate for next group.) Add much H_2O to ppt. Boil. Repeat with fresh portions of H_2O. 2..............Filter. Wash ppt. thoroughly. Add H_2SO_4 to filtrate. White ppt. = Pb_2SO_4.
Lead.	Chloride of lead is soluble in boiling water, the whole of the Lead may be dissolved out from the precipitate by boiling with successive portions of water, and separated by decantation or filtration. Test for Lead, in the filtrate or decanted liquid, with sulphuric acid.	
Silver.	Chloride of silver is soluble in hydrate of ammonium. Silver may be removed from the residue by treating with hydrate of ammonium and filtration. (If Lead is present and has not been detected, some turbidity will occur in the filtered solution, from formation of a basic salt of lead. It will dissolve in nitric acid, and therefore may be neglected.) Test for Silver, in the filtered solution, with nitric acid.	Residue from 2. (Ag. Hg.) Add NH_4HO to ppt. Heat. 3..............Filter. Add HNO_3 to filtrate. White ppt. = $AgNO_3$.
Mercury.	Any residue will consist of the black compound formed by hydrate of ammonium and mercurous chloride. Test for Mercury, by fusing the residue with carbonate of sodium.	Residue from 3. Dry the residue. Heat in tube with Na_2CO_3. Grey sublimate = Hg.

METHOD OF ANALYSIS FOR GROUP I. (CONTINUED).

TREATMENT OF AN *ALKALINE* SOLUTION WITH HYDROCHLORIC ACID.

I. The addition of hydrochloric acid to an alkaline solution may produce *a precipitate which dissolves in excess of the precipitant.*

The solution in this case is treated with hydrosulphuric acid gas, as in the regular course, according to the method for Group II.

II. A precipitate may form, on adding hydrochloric acid, *which is not dissolved in excess of the precipitant, or on boiling.*

The formation of this precipitate depends on the fact that certain salts, insoluble in water and hydrochloric acid, are held in solution by some alkali or alkaline salt. The latter is decomposed by hydrochloric acid, and the salt held in solution separates, thus:

$$\underbrace{Sb_2S_3 + (NH_4)_2S}_{\text{solution}} + 2HCl = \underbrace{Sb_2S_3}_{\text{ppt.}} + \underbrace{2NH_4Cl}_{\text{solution}} + \underbrace{H_2S}_{\text{gas}}$$

(See Gr. II. sect. I. col. 1).

ANALYSIS OF PRECIPITATE PRODUCED BY HYDROCHLORIC ACID IN AN ALKALINE SOLUTION, AND INSOLUBLE IN EXCESS.

α. Ppt. without evolution of gas.

a. *White.*

$PbCl$, Pb_2SO_4, $AgCl$, $HSiO_3$

Test for $HSiO_3$ separately. Test for the others as substances insoluble in water or acids. See Appendix to Part IV.

b. *Orange or Yellow.*

As_2S_3, As_2S_5, Sb_2S_3, Sb_2S_5, SnS_2, $HSiO_3$

1. Filter. Boil, with concd. $HCl + HNO_3$.
2. Filter (if necessary). Treat *the solution* with H_2S, as in Gr. II. If any *residue* from 2 treat as a substance insoluble in water or acids. See Appendix to Part IV.

β. Ppt. with evolution of H_2S gas alone.

a. *White.* Due to sulphur, from presence of an alkaline sulphide. Boil, and Filter.

- Ppt. Treat as a substance insoluble in water or acids. See Appendix to Part IV.
- Filtrate. Treat with NH_4HO as in Gr. III. (p. 74).

b. *Coloured.*

PtS_2, AuS_3, SnS_2, Sb_2S_3, Sb_2S_5, As_2S_3, As_2S_5, Hg_2S, Cu_2S, Ni_2S

Filter.

- Ppt. Treat with H_2S as in Group II., *if sol. in acids;* otherwise as a substance insoluble in water or acids. See Appendix to Part IV.
- Filtrate. Treat with NH_4HO as in Gr. III. (p. 74.)

γ. Ppt. with evolution of HCy gas, alone or with H_2S gas.

Presence of an alkaline cyanide or sulphide.

Boil with more HCl or HNO_3, to expel HCy. Filter.

- Ppt. Treat as a substance insoluble in water or acids. See Appendix to Part IV.
- Filtrate. Treat with H_2S gas as in Gr. II. (p. 70.)

III. No precipitate forms, *but gas is evolved,* on adding hydrochloric acid.

α. The gas smells of H_2S, and blackens a solution of acetate of lead.

Presence of an alkaline sulphide.

Boil to expel all H_2S. Treat with NH_4HO as in Gr. III. (p. 74.)

β. The gas smells of HCy.

Presence of an alkaline cyanide.

Boil to expel all HCy. Treat with H_2S as in Gr. II. (p. 70.) (It will not matter if H_2S or CO_2 are evolved along with HCy.)

γ. The gas is inodorous, and precipitates lime-water.

Presence of carbonic acid combined with an alkali.

Treat with H_2S as in Gr. II. (p. 70.)

METHOD OF ANALYSIS FOR GROUP II.

HYDROSULPHURIC ACID

precipitates from a solution in which hydrochloric acid has failed to produce a precipitate, or from the filtrate from the hydrochloric acid precipitate,

Mercuric Sulphide.

Sulphides of {Bismuth, Cadmium, Copper.}

Sulphide of Lead.

Sulphides of {Platinum, Gold.}

{Stannous and Stannic, Antimonious and Antimonic, Arsenious and Arsenic,} Sulphides.

Hydrosulphuric acid is applied in the form of gas, and is passed through the solution for a considerable time, heating at intervals, in order to ensure a complete result.

Before passing the gas, it must be ascertained that the solution has been acidified with hydrochloric acid, otherwise members of the third or fourth group might be precipitated.

Separation of Sulphur may result from the action of hydrosulphuric acid, due to the reduction of some easily reducible salt of a basic radical which is not precipitated as sulphide by the group-reagent.

A copious yellow deposit of sulphur may occur if nitric acid is present.

A yellow-white precipitate of sulphur may be due to the presence of a ferric salt.

A green colour in the solution indicates the presence of a chromic salt.

In such cases the precipitate should be invariably further examined.

Arsenic is precipitated from solutions with great difficulty by hydrosulphuric acid. If it has been detected in the preliminary examination, the solution must be largely diluted with water and then fully saturated with the gas, boiling repeatedly. If zinc is present, a portion is liable to be precipitated along with arsenic. If arsenic exists in the form of its arsen*ic* compounds, the addition of sulphurous acid before passing hydrosulphuric acid gas will, by reducing an arsen*ic* to an arseni*ous* salt, ensure complete precipitation. But if lead, barium or strontium are present, sulphurous acid will, from oxidation, cause these to be precipitated as insoluble sulphates. Hence, if the precipitate produced by hydrosulphuric acid gas, after adding sulphurous acid, is not soluble in acids, it must be treated as an insoluble substance, and examined accordingly.

EXAMINATION FOR PLATINUM AND GOLD.

If one or both of these have been detected in the preliminary examination, they must be sought for specially in a portion of the solution in which hydrochloric acid has failed to produce a precipitate or in the filtrate from the hydrochloric acid precipitate.

Their reactions are so characteristic that these metals will be detected in presence of all other basic radicals.

All nitric acid, if present, must be removed by one or two evaporations with hydrochloric acid.

Platinum. Platinum is converted into chloro-platinate by means of chloride of ammonium, and separated by filtration.

Gold. Gold may be detected in the residual solution by ferrous sulphate or oxalic acid.

N.B. This process will occupy some hours to be efficiently performed, as the platinum precipitate forms slowly.

Add to a portion of the original solution
$HCl + NH_4Cl$ or KCl,
(After expelling all HNO_3).
Evaporate to dryness.
Treat residue with alcohol.
Yellow ppt. = NH_4PtCl_6 or $KPtCl_6$.
Filter.

Filtrate.
Evaporate to expel alcohol.
Add Fe_2SO_4 to residue.
Brown powder = Au.

METHOD OF ANALYSIS FOR GROUP II. (CONTINUED).

ANALYSIS OF PRECIPITATE PRODUCED BY HYDROSULPHURIC ACID IN AN ACID SOLUTION.

The *filtrate*, which may contain members of other groups, is set aside for further examination.

The *precipitate* after being well washed with hot water, containing some hydrosulphuric acid, is boiled with sulphide of ammonium, in order to separate this group into two sections.

PORTION OF PRECIPITATE *INSOLUBLE* IN SULPHIDE OF AMMONIUM.

Mercury.

The insoluble residue is washed thoroughly free from all sulphide of ammonium and boiled with nitric acid. Sulphide of mercury remains as an insoluble residue. *Before* filtering this from the undissolved portion, sulphuric acid is added, to precipitate any Lead in the form of sulphate, and this removed *together with* the mercuric sulphide. The insoluble residue is a mixture of Hg_2S and Pb_2SO_4, but neither will interfere with the detection of the other.

A portion of the mixed precipitate is dissolved in nitro-hydrochloric acid, and from this solution Mercury may be obtained as a metallic deposit on the surface of a piece of copper foil.

[Lead.]

Another portion is ignited on platinum foil or porcelain. Sulphate of lead not being volatile will remain as a white residue upon the foil or porcelain.

Bismuth.

The remaining sulphides (Bismuth, Cadmium and Copper) in solution in nitric acid, are treated with hydrate of ammonium in excess and heat applied. Bismuth is alone precipitated from the solution as hydrate.

Test by dissolving in very little hydrochloric acid and adding much water.

Cadmium and Copper. (and see next page.)

The solution containing Cadmium and Copper is evaporated nearly to dryness, dissolved in acetic acid and water, and the metals again converted into sulphides by passing hydrosulphuric acid through the solution.

Sulphide of copper may then be dissolved by adding cyanide of potassium and separated from the insoluble sulphide of cadmium, which remains as a yellow residue.

Test for Cadmium by blowpipe. For Copper, by means of ferro-cyanide of potassium, previously adding acetic acid, since other acids if present decompose the precipitate formed.

To the solution to be analysed add H_2S gas.
(Observing precautions.)
1.............Filter.
(Examine filtrate for next group.
Wash ppt. thoroughly with H_2S water.
Boil ppt. with $(NH_4)_2S$.
2.............Filter a.

Residue from 2. (Hg. Pb. Bi. Cd. Cu.)
Wash thoroughly.
Boil with HNO_3, until all red fumes cease.
Dilute with H_2O.
Add H_2SO_4 until no more ppt. forms.
3.............Filter.
Divide *residue* into two portions, A and B.
A. Dissolve in $HCl + HNO_3$.
Introduce piece of metallic copper.
Metallic coating = Hg.
B. Ignite on platinum foil or porcelain.
White residue = Pb_2SO_4.

Filtrate from 3. (Bi. Cd. Cu.)
Add NH_4HO in excess.
Heat.
4.............Filter.
Dry *the ppt.* slightly.
Dissolve in *very little* HCl, add H_2O.
White ppt. = $BiCl_3 . Bi_2O_3$.

Filtrate from 4. (Cd. Cu.)
Evaporate nearly to dryness.
Add $HC_2H_3O_2$ and H_2O.
Reppt. by H_2S gas.
Decant the liquid.
Add *to ppt.* KCy.
5.............Filter.
Yellow residue = Cd_2S.
Confirm by blowpipe.

Filtrate from 5. (Cu.)
Add some acetic acid.
Add K_2Cfy.
Brown ppt. = Cu_2Cfy.

---

a The filtrate, which may contain Sb, As or Sn, is treated as on p. 72, or by Hofmann's method, p. 73.

METHOD OF ANALYSIS FOR GROUP II. (CONTINUED).

METHOD FOR THE SEPARATION OF CADMIUM AND COPPER (HOFMANN).

A solution, which may contain both Cadmium and Copper, is acidified (if not already acid) with hydrochloric acid.

Hydrosulphuric acid gas is then passed through the solution, to precipitate the metals as sulphides.

The precipitate must be washed thoroughly and *quickly*, otherwise sulphide of copper might pass by oxidation into cupric sulphate.

Cadmium. The precipitate is boiled with dilute sulphuric acid. Sulphide of cadmium is thus decomposed, and soluble sulphate of cadmium formed.

Sulphide of copper remains undissolved.

Test for Cadmium in the solution by hydrosulphuric acid.

Copper. The residue is dissolved in nitric acid, excess of hydrate of ammonium added to neutralize the acid, and, finally, excess of acetic acid to prevent decomposition of the precipitate produced by ferrocyanide of potassium.

Test for Copper in the acetic acid solution by ferrocyanide of potassium.

To solution to be analysed, add HCl.
(If not already acid).
Pass H_2S gas.
Wash ppt. thoroughly and *quickly*, by decantation.
Add H_2SO_4.
Boil.
1............Filter.
Pass H_2S gas through *filtrate*.
Yellow ppt. = Cd_2S.

Residue from 1. Cu.
Dissolve in HNO_3.
Add NH_4HO in excess.
Add $HC_2H_3O_2$ in excess.
Add K_4Cfy.
Red-brown ppt. = Cu_2Cfy.

N.B. All the operations in the above method must be performed as rapidly as possible.

PORTION OF PRECIPITATE *SOLUBLE* IN SULPHIDE OF AMMONIUM.

The sulphides of Tin, Antimony, and Arsenic, held in solution by sulphide of ammonium, are reprecipitated as sulphides on the addition of hydrochloric acid.

As hydrochloric acid is liable however to produce decomposition and solution as chloride, of a portion of the sulphides, it is advisable to pass some hydrosulphuric acid gas through the solution at the same time. The precipitate by hydrochloric acid must be well washed.

Arsenic. The reprecipitated sulphides are treated with carbonate of ammonium, which dissolves only sulphide of arsenic.

Arsenic is again precipitated from this solution by hydrochloric acid, and tested by Fresenius' and Babo's test, or the solution may be precipitated by hydrochloric acid, the precipitate dissolved in nitro-hydrochloric acid, and tested by Marsh's test (p. 27).

Antimony. The undissolved sulphides of antimony and tin, separated by carbonate of ammonium from arsenic, may be dissolved in nitro-hydrochloric acid.

Antimony is sought for by means of a Marsh's apparatus (p. 27).

Tin. Tin is reduced by zinc in the apparatus at the same time as a black metallic powder. This, if collected, washed, dissolved in hydrochloric acid, and treated with mercuric chloride, yields the characteristic grey precipitate (p. 24).

After boiling the H_2S ppt. with $(NH_4)_2S$, reppt. by HCl, passing some H_2S gas.
Wash the ppt. thoroughly.
(Wash by filtration if much ppt. of sulphur.)
Add $(NH_4)_2CO_3$.
1............Filter.
Arsenic is in the filtrate.
Add HCl *to filtrate*.
Test ppt. by Fresenius' and Babo's test (p. 29).
Or
dissolve ppt. in HCl + HNO_3.
Test by Marsh's test (p. 27).

Residue from 1. (Sb. Sn.)
Add HCl + HNO_3.
Place solution in a Marsh's apparatus.
Antimony if present will be recognised by its reactions.

Tin reduced by zinc in the apparatus, as a *Black metallic powder*.
Wash the powder.
Dissolve in HCl.
Add $HgCl$.
Grey ppt. = Hg_2Cl.

METHOD OF ANALYSIS FOR GROUP II. (CONTINUED).

SEPARATION OF ANTIMONY, ARSENIC, AND TIN (HOFMANN).

A solution supposed to contain either Antimony, **Arsenic**, or Tin, or all of them, is introduced into the generator of a hydrogen **apparatus**.

(Hydrogen is evolved by the action of sulphuric acid upon zinc, care being taken to have the materials perfectly free from arsenic.)

The evolved gases are allowed to pass from the generator, through a solution of acetate of **lead, in order** to absorb any hydrochloric or hydrosulphuric acid.

Thence they are conducted through a solution of nitrate of silver.

Antimony is here *precipitated* as antimonide of silver (Ag_3Sb), whilst Arsenic, first converted into arsenious acid, is *held in solution* as arsenite of silver (Ag_3AsO_3) by the liberated nitric acid.

$3AgNO_3 + H_3AsO_3 = Ag_3AsO_3 + 3HNO_3$ (p. 25).

Arsenic. **If *the solution* is exactly neutralised by hydrate of ammonium, arsenite of silver is precipitated (yellow).**

Antimony. *The precipitate* of antimonide of silver may be **dissolved** in boiling tartaric acid, the solution acidified by nitric acid, and hydrosulphuric acid gas passed through it. An orange precipitate of antimonious sulphide indicates the presence of Antimony.

Tin. **Tin remains** in the generator as a black deposit upon the zinc, together **with some Lead** and Antimony (these latter may be neglected). If collected and **boiled with hydrochloric acid, chloride** of tin is formed. This yields **with mercuric chloride the characteristic grey** precipitate of **mercurous** chloride (p. 24).

If the filtrate from 2 (p. 71) is used, Sb, As and Sn will be in the form of acid radicals.
Reppt. as sulphides by HCl, passing at same time some H_2S gas.
Collect the ppt. and wash it.
Boil the ppt. with $HCl + HNO_3$.
The solution may be tested as follows.

Place the solution **in a H** apparatus,
(Taking precautions in its use).
Pass evolved gases through $PbC_2H_3O_2$ solution.
Thence through $AgNO_3$ solution, until no ppt. forms in the latter.

1......Filter the $AgNO_3$ solution.
Add NH_4HO *to filtrate*, to neutralize exactly.
Yellow ppt. = Ag_3AsO_3.

Residue from 1. (Sb. Sn.)
Boil with tartaric acid.
2..............Filter.
Add HNO_3 *to filtrate*.
Pass H_2S gas.
Orange ppt. = Sb_2S_3.
(The residue from 2 consists of Ag.)

Collect **the black deposit on** Zn **in the generator.**
Boil with HCl.
Filter.
Dilute solution with H_2O.
Add **a few** drops of $HgCl_2$.
Grey ppt. = Hg_2Cl_2.
(Presence of Tin.)

N.B. It will be necessary in all cases where Tin, Antimony, or Arsenic is detected in the course of analysis, to examine a portion of the *original solution* according to reactions on pages 24 and 25, to ascertain in what state of oxidation these basic radicals may have been originally present.

METHOD OF ANALYSIS FOR GROUP III.

HYDRATE OF AMMONIUM,

in the presence of chloride of ammonium, precipitates from a solution in which hydrochloric acid and hydrosulphuric acid have failed to produce a precipitate, or in the filtrate from the hydrosulphuric acid precipitate,

Aluminium, Chromium, Iron,	as hydrates. And as phosphates in presence of the phosphoric radical.	Barium, Strontium, Calcium,	as phosphates (borates, fluorides, or oxalates) in presence of the corresponding acid radicals.
		Magnesium	as phosphate, in presence of the phosphoric radical.

A small quantity of Manganese may be precipitated with Iron, but will not interfere with the detection of the latter.

A portion of the original solution should be tested, before adding the group-reagent, *if the presence of Iron is suspected*, in order to ascertain in what state of oxidation it exists. The reactions of ferrous and ferric salts with ferro- and ferri- cyanide of potassium respectively, will determine this (p. 33).

The filtrate from the hydrosulphuric acid precipitate, or the solution in which that reagent has failed to produce a precipitate, *must be thoroughly boiled to expel all trace of the gas*, before adding hydrate of ammonium. Otherwise, on subsequently adding nitric acid (to peroxidise Iron), oxidation of sulphur might produce sulphuric acid, and this would precipitate any Barium, Strontium, or possibly Calcium, as insoluble sulphates.

Nitric acid is then added to the solution to be tested, *the whole evaporated to dryness and then ignited*. This may be repeated. By this means any Iron present is converted into a ferric salt (even if it were originally present as a ferric salt, hydrosulphuric acid will have reduced it), and hence it will be thrown down by hydrate of ammonium (p. 33).

Any oxalic acid is at the same time oxidised, and the presence of Barium, Strontium or Calcium as oxalates in the hydrate of ammonium precipitate, is prevented.

Any silica is also converted by the desiccation into insoluble silicic anhydride (Si_2O_4). If allowed to remain in its soluble form, it might be mistaken for hydrate of aluminium. (The residue after evaporation should be ignited *slightly*, not too strongly, otherwise certain oxides of this group might be rendered insoluble.)

The dry residue is dissolved in hydrochloric acid.

Chloride of ammonium is added to the solution in hydrochloric acid. This prevents precipitation of Magnesium (except as phosphate in presence of the phosphoric radical) by hydrate of ammonium.

Manganese may be detected in the hydrate of ammonium precipitate, by fusing a small portion with carbonate of sodium and cyanide or nitrate of potassium on platinum foil; a bluish-green residue indicates its presence.

The phosphoric radical may also be detected in a small portion of the precipitate, by dissolving in nitric acid, and adding nitrate of silver. Its presence is revealed by a yellow ppt. The absence of this radical will necessitate great abridgement in the following method of analysis.

Borates and fluorides may be neglected, as sufficient both of the basic and acid radicals will remain to be detected in the appropriate place.

METHOD OF ANALYSIS FOR GROUP III. (CONTINUED).

ANALYSIS OF PRECIPITATE PRODUCED BY HYDRATE OF AMMONIUM.

The *filtrate*, which may contain members of other groups, is set aside for further examination.

The *precipitate* must be thoroughly washed until quite free from all trace of ammonia. It is dissolved in dilute hydrochloric or nitric acid.

Excess of hydrate of potassium added in the cold will then precipitate Iron (as phosphate or hydrate) together with the phosphates of barium, strontium, or calcium. Chromium and Aluminium remaining in solution.

Chromium. By long-continued boiling of the solution, Chromium is precipitated, and Aluminium left in solution.

As some Iron, if present, may have passed into solution along with Chromium, it is necessary to examine the precipitate produced on boiling, by fusing a portion with carbonate of sodium and nitrate of potassium. Chromium, converted into soluble chromate of sodium, may be dissolved out by water and tested with acetate of lead. Iron will be detected in the residue. The remaining portion of the precipitate is examined with a view to ascertain whether Chromium existed originally in combination with the phosphoric radical, or otherwise.

Aluminium. In the solution from which Chromium has been separated by boiling, Aluminium may exist as hydrate or phosphate. The solution is treated with acetic acid in excess.

Phosphate of aluminium, if present, being insoluble in acetic acid, will be precipitated.

Test for Aluminium, in the solution, with hydrate of ammonium. After removing Aluminium from the precipitate of phosphate of aluminium, by means of silicate of potassium and filtration, the solution may be tested for the phosphoric radical with molybdate of ammonium (p. 46).

Iron. The precipitate containing Iron (together possibly with some Manganese), and the phosphates of the alkaline earths, is dissolved in hydrochloric acid. Acetate of potassium added in excess precipitates Iron only, as phosphate (if the phosphoric radical is present), and as basic acetate. A portion of this precipitate should be tested for Chromium, as directed above, unless the latter has been already detected. The other portion is treated with acetic acid.

Any residue after adding acetic acid will contain the phosphoric radical, phosphate of iron being insoluble in acetic acid. Test *the solution* with ferrocyanide of potassium.

If the solution to be analysed has been treated previously with H_2S,
boil,
to expel all trace of H_2S.
Add HNO_3,
and evaporate to dryness once or twice.
Dissolve residue in HCl.
Add some NH_4Cl.
Then add group reagent, NH_4HO.

1............Filter.
(Examine *filtrate* for next group.)
Wash *the ppt.* thoroughly.
Dissolve in HCl,
Add *in the cold*, KHO in excess.

2............Filter.
Boil *the filtrate* for some time.

3............Filter.
Divide *ppt.* into two portions, A and B.

A. Fuse with $Na_2CO_3+KNO_3$.
Digest with H_2O.
Dissolve *residue* in $HC_2H_3O_2$, and test for Fe, with K_4Cfy
To solution add $HC_2H_3O_2$, and $PbC_2H_3O_2$.
Yellow ppt. = $PbCrO_4$.

B. Dissolve in HCl.
Add $KC_2H_3O_2$ in excess.
Ppt. Indicates the phosphate of chromium.
Solution. Indicates the oxide. Confirm by NH_4HO.

Filtrate from 3. Al.
Add $HC_2H_3O_2$ in excess.
Dissolve *any ppt. so formed*, in HCl.
Add KHO in excess.
Add $KSiO_2$.
White ppt. = $Al_2(SiO_3)_3$.
(The solution may be tested for the phosphoric radical with molybdate of ammonium as in V. p. 46.)
The solution after adding $HC_2H_3O_2$ may be tested for Al by NH_4HO.

Residue from 2.
[Fe, (Ba, Sr, Ca.)]
Dissolve in HCl.
Add KC_2H_3O in excess.

4............Filter.
(A portion of ppt. is fused for Cr. as above.)
Add to *ppt.* $HC_2H_3O_2$.
Add K_4Cfy.
Blue ppt. = $(Fe_2)_2Cfy$.
(Any residue after adding $HC_2H_3O_2$ consists of phosphate of iron.)

Filtrate from 4.
Examine for Ba, Sr, and Ca, as on page 77.

METHOD OF ANALYSIS FOR GROUP IV.

SULPHIDE OF AMMONIUM

precipitates from a solution in which hydrochloric acid, hydrosulphuric acid and hydrate of ammonium have failed to produce a precipitate, or from the filtrate from the hydrate of ammonium precipitate,

Sulphides of { Manganese, Zinc, Cobalt, Nickel.

On filtering the precipitate produced by sulphide of ammonium, the filtrate may appear dark coloured. This is due to the presence of sulphide of nickel, which is soluble to some extent in sulphide of ammonium (Nickel, p. 35, note β). In this case the sulphide of nickel may be separated from the filtrate by heat, or by evaporating to expel all sulphide of ammonium, and adding hydrochloric acid. The precipitate so formed should be examined with that thrown down by the group reagent.

ANALYSIS OF PRECIPITATE PRODUCED BY SULPHIDE OF AMMONIUM.

The *filtrate*, which may contain members of other groups, is set aside for further examination. Some difficulty may be experienced in obtaining a *clear* filtrate; repeated filtration can alone effect a satisfactory result.

The *precipitate* must be well washed with water, and some sulphide of ammonium repeatedly added at the same time.

Hydrochloric acid is employed to dissolve the sulphides of manganese and zinc, and to separate them from the sulphides of cobalt and nickel.

Manganese. Manganese may then be precipitated from the hydrochloric acid solution by means of *excess* of hydrate of potassium.

Test for Manganese by blowpipe, or sulphide of ammonium, in the latter case previously dissolving in an acid (except acetic acid).

Zinc. Test for Zinc, in the solution, by hydrosulphuric acid in presence of acetic acid (p. 34).

Cobalt and Nickel. The remaining sulphides of cobalt and nickel may be dissolved in nitro-hydrochloric acid, care being taken to employ only just sufficient nitric acid to effect the solution.

The metals may be converted into cyanides, and held in solution, by the addition of cyanide of potassium in excess. On boiling the solution with a few drops of hydrochloric acid, cobalticyanide of potassium and cyanide of nickel and potassium are produced (see p. 35, note δ). Whilst still boiling, if a solution of hypochlorite of sodium *in excess* is added, the Potassium and Cyanogen are taken up and Nickel is precipitated as black sesquioxide. In this state it may be readily removed from the Cobalt which remains in solution.

(Chlorine and mercuric nitrate have also been employed instead of hypochlorite of sodium.)

To the solution to be analysed, add $(NH_4)_2S$.
1............Filter.
(Examine filtrate for next group. *If the filtrate is dark coloured*, evaporate. Add HCl. Examine the ppt. so formed with that from 1.)
Wash *ppt. from* 1 thoroughly.
Add HCl in the cold.
2............Filter.
Boil *the filtrate* to expel all H_2S.
Add KHO in excess.
Whitish ppt. = MnHO.
Confirm by blowpipe or $(NH_4)_2S$.
3............Filter.

Filtrate from 3. (Zn.)
Add $HC_2H_3O_2$.
Pass H_2S gas.
White ppt. = $Zn_2S.H_2O$.

Residue from 2. (Co, Ni.)
Test a portion with borax for Co.
Add HCl.
Boil.
Add few drops of HNO_3, just to dissolve.
Add KCy in excess, and few drops of HCl.
Boil.
Add NaClO in excess, to boiling solution.
Black ppt. = $(Ni_2)_2O_3$.
4............Filter.

Filtrate from 4. (Co.)
Evaporate to dryness.
Heat with borax on platinum wire in the blowpipe flame.
Blue bead indicates Co.

METHOD OF ANALYSIS FOR GROUP V.

CARBONATE OF AMMONIUM,

in the presence of chloride of ammonium, precipitates from a solution in which hydrochloric acid, hydrosulphuric acid, hydrate of ammonium and sulphide of ammonium have failed to produce a precipitate, or from the filtrate from the sulphide of ammonium precipitate,

Carbonates of { Barium, Strontium, Calcium.

Chloride of Ammonium is added, as in the case of Group III., to prevent the precipitation of Magnesium, as it would not be *completely* thrown down by carbonate of ammonium.

Hydrate of Ammonium is added, previously to the group-reagent, to ensure against an acid solution.

The precipitation will be facilitated by the application of heat.

ANALYSIS OF PRECIPITATE PRODUCED BY CARBONATE OF AMMONIUM.

	The *filtrate*, which may contain members of the sixth group, is set aside for further examination.	If the solution to be analysed has been treated with $(NH_4)_2S$, all the latter must be boiled off, and the solution filtered from any ppt. of S. Add NH_4Cl. Add NH_4HO. Then add group reagent $(NH_4)_2CO_3$.
Barium.	The *precipitate* is washed, dissolved in hydrochloric acid, and Barium separated as fluosilicate by fluosilicic acid and alcohol. The whole of the Barium may be removed by taking care to evaporate sufficiently with alcohol and to allow time for the separation.	1............Filter. (Examine filtrate for next group.) Dissolve *the ppt.* in HCl. Add $H_2Si_2F_6$ and agitate. Evaporate to dryness, digesting with alcohol and H_2O. **White ppt. = $Ba_2Si_2F_6$.** 2............Filter.
Calcium.	Ferrocyanide of potassium will precipitate Calcium from the remaining solution, on boiling.	Solution from 2. (Ca, Sr.) Evaporate with more $H_2Si_2F_6$ and alcohol, to dryness. Treat the residue with water. Filter, from any ppt. of $Ba_2Si_2F_6$. Add to *solution* K_4Cfy. Boil. **White ppt. = Ca_2Cfy.** 3............Filter.
Strontium.	Test for Strontium, in the *solution*, by sulphate of calcium or oxalic acid, or in the solution evaporated to dryness, by blowpipe.	Filtrate from 3. Evaporate to dryness. Test on platinum foil before blow-pipe. **Crimson flame indicates Sr.** (Strontium may also be precipitated from the filtrate from 3, by sulphate of calcium or oxalic acid.)

METHOD OF ANALYSIS FOR GROUP VI.

NO PRECIPITATE

is produced in solutions containing the members of this group, by the reagents which are employed to precipitate members of the other groups.

A solution in which the previous group-reagents have failed to produce a precipitate, or the filtrate from the carbonate of ammonium precipitate, may contain

Magnesium,
Ammonium,
Potassium,
Sodium, } combined with one or more acid radicals.

Ammonium will have been already detected, if present, in the preliminary examination. If detected, it must be expelled before proceeding to test for other members of the group.

The behaviour of Ammonium and Potassium with those reagents employed to detect the latter are so similar as to render this precaution essential.

ANALYSIS OF A SOLUTION WHICH MAY CONTAIN THE METALS, MAGNESIUM, POTASSIUM, SODIUM, AND (AMMONIUM).

	The solution is divided into two portions, A and B.	Divide solution into two portions, A and B.
Magnesium.	A. Magnesium may be detected in this portion by phosphate of sodium in presence of hydrate of ammonium. Time must be allowed for the separation of the Magnesium salt, as the precipitate always forms with difficulty.	A. Add NH_4HO, and Na_2HPO_4. Agitate well and allow to stand some time. **White ppt. = $MgNH_4PO_4$.**
Ammonium.	B. This portion of the solution is evaporated to dryness, and the residue ignited, until all fuming ceases, to completely expel ammonia. The characteristic reaction with hydrochloric acid will determine this, and at the same time identify the presence of ammonium.	B. Evaporate to dryness. Ignite, until all fuming ceases. (Test fumes with HCl on a glass rod.) Add to residue H_2O.
Potassium.	The residue is treated with a small quantity of water. Potassium is converted into chloro-platinate by the addition of hydro-chloro-platinic acid (bi-chloride of platinum) to the solution, together with some alcohol to promote precipitation.	1............Filter (if necessary). Add to *solution* HCl and $HPtCl_6$, and some alcohol. **Yellow ppt. = $KPtCl_6$.** 2............Filter.
Sodium.	The remaining solution contains the Sodium salt. The solution may be evaporated to dryness, and Sodium tested for in the residue, before the blowpipe. Care must be taken to have the platinum wire or foil *perfectly* clean, as even moisture from the fingers may impair the delicacy of the reaction.	Filtrate from 2. Evaporate to dryness. Test on platinum wire before blowpipe. **Yellow flame indicates Na.**

PART IV.

SYSTEMATIC COURSE OF ANALYSIS

FOR

ACID RADICALS.

ANALYSIS OF A SALT OR MIXTURE OF SALTS FOR ACID RADICALS.

The Course of Analysis (as in the case of basic radicals) is divided into three stages, or series of processes.

Many of the rules laid down, as affecting the systematic course for basic radicals (pages 60 and 61), are no less applicable to and important in the present course for acid radicals.

Certain special features demand attention, as an introduction to the course detailed in this Part.

I. The preliminary examination. (The substance under analysis being in the solid form.) p. 82.

The decompositions effected on submitting a body to a high temperature, either alone or with certain concentrated acids, chiefly hydrochloric acid, nitric acid, or sulphuric acid, and the characteristic colour, odour or reactions of the substances which result from such decomposition, are relied upon to detect the presence of individual acid radicals.

More importance is attached to the results of a preliminary examination, as serving to reveal the presence or absence of acid radicals, than is assigned to such an examination in the case of basic radicals. This arises from the nature of the reactions of acid radicals with reagents in solution.

The presence or otherwise of certain acid radicals will be determined by the cautious observer during the preliminary examination for basic radicals.

III. The process of solution.

The explanations already offered upon this part of the analytical course apply in this place. When it is necessary to resort to the action of an acid in order to effect solution, it will usually be found most advantageous to employ nitric acid (as this acid forms no insoluble salts).

III. The application of reagents to the solution.

The first thing which claims attention under this head is the preparation of the solution to be examined. This preparation becomes necessary, since if certain basic or acid radicals are present in a substance under analysis, and are allowed to remain in a solution, the application of certain reagents would lead to results different from those which would be anticipated. Thus, the presence of *carbonic acid* might lead to the formation of a precipitate, where this result would not only *not* lead to the detection of the carbonic radical itself, but falsify other results which are looked for. In like manner the presence of any basic radicals, *except potassium, sodium,* or *ammonium,* would frequently cause precipitation in a solution, instead of neutralisation merely, when hydrate of ammonium was employed as a reagent. Ammonium, again, must be excluded, in consequence of the solvent action which many of its combinations, especially the chloride, exercise upon salts which it may be desirable to obtain in an insoluble form.

In the examination for acid radicals the reagents employed do not effect *separation* of the acid radicals, to the same extent as was the case with basic radicals. The reagents only serve to *identify,* as a rule, the presence or absence of individual radicals. The results of the application of *general reagents* may be regarded as affording so many hints or clues, determining merely the selection of a series of special and confirmatory tests. It would be useless therefore to apply a general reagent to the *filtrate* from a precipitate thrown down by another general reagent, in a manner similar to that adopted in analysing for basic radicals. The three general reagents, chloride of barium, nitrate of silver, and ferric chloride, employed in the present course, are to be applied to *fresh portions of the solution, prepared as directed on page* 83. In some instances even the use of a solution prepared in such a manner is proscribed; some special tests are applied to the *original solution*. It is clear that great economy in the use of material is essential, where the solution to be analysed has to be split up into so many separate portions.

The methods for the analysis of acid radicals, given in the following pages, are subject to many modifications dependent on the results of previous examinations. These results are chiefly—the nature of the basic radicals found in the solution—the solubility of various salts (an aqueous solution which contains barium would not be examined for sulphuric acid, as sulphate of barium is insoluble; a solution in which silver has been detected could not contain hydrochloric acid)—the reactions of a solution with vegetable colours.

Thus it happens that much forethought, caution, and judgment are requisite on the part of the analyst who wishes to attain success without sacrifice of time and material, when an examination for acid radicals is undertaken.

PRELIMINARY EXAMINATION.

(ACID RADICALS.)

Exp. I. A portion of the powdered substance is heated in a **Glass tube open at both ends,** and held obliquely in the flame.	Results of this experiment, as in Exp. I. Basic radicals, page 62.	
Exp. II. i. A portion of the powdered substance is treated with 3 or 4 times its bulk of **Dilute Hydrochloric acid,** avoiding heat.	An odorous, yellow gas is evolved. The gas explodes in contact with flame. (*the reactions in* ii. *may also be observed here.*)	Chlorate.
ii. The above mixture is heated (only cautiously if a chlorate is present).	Effervescence. The escaping gas is *inodorous*, and precipitates lime-water.	Carbonate.
	Odour of hydrocyanic acid.	Cyanide. Ferric ide.
	Odour of hydrosulphuric acid gas.	Sulphide of an al alkaline earth, manganese or
	Odour of burning sulphur.	Sulphite. Hyp phite (or from oxidation of sul ric acid).
	Evolution of red fumes (in presence of copper).	Nitrate.
Exp. III. A portion of the powdered substance is heated, nearly to boiling, with 3 or 4 times its bulk of **Concentrated Sulphuric acid.**	*No blackening of the mixture.* Effervescence, with evolution of gases. Gas burns with a blue flame.	Oxalate. Cyani Ferricyanide. Ferrocyanide. phocyanide.
	has odour of burning sulphur.	Sulphite. Hyp phite (or from oxidation of sul ric acid).
	gives white fumes in contact with hydrate of am. on a glass rod.	Chloride. (excep Hg, Ag, Pb Sn.)
	has odour of hydrosulphuric acid gas.	Sulphide.
	has an aromatic odour.	Acetate.
	Coloured vapours are evolved. Red-brown; pungent odour; colour starch-paste, *orange.*	Bromide.
	Violet; colour starch-paste, *blue.*	Iodide.
	Greenish-yellow; detonate.	Chlorate.
	Red.	Nitrate.
	Colourless gases are evolved. Gas kindles burning wood.	Chromate.
	Pungent fumes are evolved. Fumes give film of silicic acid on wet glass rod.	Fluoride.
	Crystals, as scales, appear in the solution, on cooling.	Borate.
	Blackening of the mixture. Indicates presence of a non-volatile organic acid.	Tartrate. Citra Tannate, Gal
	Odour of burnt sugar; evolution of carbonic oxide which burns with blue flame.	Tartrate.

CONFIRMATORY TESTS

for acid-radicals*.

Cyanogen.	Vapours passed into solution of KHO; add Fe_2SO_4 and Fe_2Cl_6; dissolve any ppt. in HCl—*blue ppt.*
Sulphurous radical.	Gas passed into solution of bichromate of potassium—*green colour.*
Chlorine.	Substance heated with H_2SO_4+$KCrO_2$; fumes condensed; NH_4HO added to condensed liquid—*yellow colour.* Acetic acid and $AgNO_3$ added to this—*crimson ppt.*
Sulphur.	(Sulphide). Paper moistened with acetate of lead held over gas—*black ppt.*
Acetic radical.	Substance heated with H_2SO_4 + alcohol; acetic ether evolved—*aromatic odour.*
Nitric radical.	Substance + H_2SO_4 + crystal of Fe_2SO_4—*brown ring round crystal.*
Fluorine.	Substance + H_2SO_4 + sand; products conducted into water—*gelatinous ppt.*
Tannic radical.	Substance in solution + solution of gelatine—*white viscous ppt.*
Citric radical.	Substance in solution + excess of hydrate of calcium—*white ppt. on boiling, dissolving when cool.*

* It is intended that the evidence furnished by the results of Exp. III. should be followed up by employing these confirmatory tests.

PREPARATION OF THE SOLUTION WHICH IS TO BE EXAMINED FOR ACID RADICALS.

Hydrochloric acid.

If, on examining for basic radicals, the original solution was *acid or neutral*, and CARBONIC ACID was detected (by effervescence and reaction with hydrate of calcium), boil the solution to be prepared (a similar one to that used in the analysis for basic radicals) with hydrochloric acid, *in order to expel all carbonic acid.*

If carbonic acid was not detected, this step may be omitted.

If the original solution was *alkaline*, the addition of hydrochloric acid might indicate the probable presence of one or more of the following acids, if present:

CARBONIC ACID, HYDROSULPHURIC ACID, HYDROCYANIC ACID, Silicic acid, Hyposulphurous acid, Hydroferrocyanic acid, Benzoic acid.

If Silver, Lead, or Mercury were present in the original solution, these will be here precipitated by hydrochloric acid, and must be removed by filtration.

Hydrosulphuric acid.

Through the remaining part of the solution pass hydrosulphuric acid gas. The following, if present, will be decomposed and their presence rendered more or less evident:

SULPHUROUS ACID (in an acid solution), IODIC ACID (reduced to iodide), page 64.

The presence of Tin, Antimony, Arsenic, and Chromium, as acid radicals, would be revealed by the action of hydrosulphuric acid, when testing for basic radicals.

Remove, by filtration, any bodies precipitated by hydrosulphuric acid.

Sulphide of Ammonium.

If any members of Groups III. and IV. (Basic radicals) were detected during the examination for basic radicals, precipitate them by sulphide of ammonium and remove.

Carbonate of Sodium.

After boiling well, to expel all hydrosulphuric acid, add to the solution carbonate of sodium in some quantity. All basic radicals, *except Potassium, Sodium, and Ammonium*, will thus be precipitated and must be removed.

Hydrate of Potassium.

If Ammonium has been detected in the previous analysis, boil the filtrate from the carbonate of sodium precipitate, with hydrate of potassium, until no ammonia is evolved.

Test the solution, thus prepared, with test-paper.

Render it *neutral* by hydrate of ammonium or by nitric acid, as may be required, taking care to exercise extreme delicacy in bringing it into this condition.

The prepared solution may now be examined for acid radicals, according to the plan detailed in the following pages.

BEHAVIOUR OF ACID RADICALS AN

	Chloride of Barium in a neutral solution.	**Nitrate of Silver** in an acidified solution.
Group I. Sect. I.		
Sulphuric acid . . 1.	1. **Sulphate,** white . } insol. in HCl.	
Hydrofluosilicic acid *. 2.	2. **Fluosilicate,** white. }	
Sect. II.		
Carbonic acid . . 3.	3. Carbonate, white . } sol. in HCl. with decomposition.	
Silicic „ . . 4.	4. **Silicate,** white. }	4. White, from a neutral solution.
[Hydrosulphuric „]		
Sect. III.		
Phosphoric acid . . 5.	5. **Phosphate,** white . } sol. in HCl. without decomposition.	5. Yellow, from a neutral solution.
Boracic „ . . 6.	6. **Borate,** white . }	6. White, „ „
Oxalic „ . . 7.	7. **Oxalate,** white . }	7. White, „ „
Hydrofluoric „ . . 8.	8. **Fluoride,** white . }	
Sect. IV.		
Chromic acid . . 9.	9. Chromate, pale-yellow . } acid solution decomposed by hydrosulphuric acid.	9. Crimson, from a neutral solution.
Sulphurous acid . . 10.	10. Sulphite, white . }	
Hyposulphurous acid . 11.	11. Hyposulphite, white . }	
Iodic „ . 12.	12. Iodate, white . }	
[Arsenious and arsenic acids] 13.		13. (See Tables for Basic radicals.)
Group II. Sect. I.		
Hydrochloric acid . . 1.		1. **Chloride,** a. white } sol. in hydrat of am- moniun
Hydrobromic „ . . 2.		2. **Bromide,** a. white-yellow }
Hydrocyanic „ . . 3.		3. **Cyanide,** a. white }
Hydroferricyanic acid . 4.		4. **Ferricyanide,** a. orange }
Sect. II.		
Hydriodic acid . . 5.		5. **Iodide,** a. pale-yellow } insol. ir hydrate of am- monium
Hydrosulphuric acid . 6.		6. **Sulphide,** a. black }
Hydroferrocyanic acid . 7.		7. **Ferrocyanide,** a. white }
Hydrosulphocyanic „ 8.		8. **Sulphocyanide,** a. „ }
Group III.		
Nitric acid.	Not precipitated.	Not precipitated.
Chloric „		
Organic Acids.		
Group I. Tartaric acid . 1.	1. **Tartrate,** white.	1. White, from a neutral solution.
Citric „ . 2.	2. **Citrate,** white.	2. White, „ „
Group II. Succinic acid . 1.		
Benzoic „ . 2.		
Tannic „ . 3.	3. Tannate, white.	
Gallic „ . . 4.	4. Gallate, white (only partially).	
Group III. Acetic acid . . 1.		
Uric „ . 2	2. Urate.	

[T]HEIR SALTS WITH GENERAL REAGENTS.

Chloride of Calcium.	Ferric Chloride in a neutral solution.	Detected along with Basic radicals. Set free with effervescence by *Hydrochloric acid*.	Decomposed by *Hydrosulphuric acid* in an acid solution.
5. **Phosphate,** a. white }sol. in acetic acid. 6. **Borate,** a. white 7. **Oxalate,** a. white }insol. in acetic acid. 8. **Fluoride,** a. white	5. White.	3. Ppt. with CaHO.	9. { Green **solution with separation of sulphur.** 10. Separation **of sulphur.** 11. " " 12. Reduction to Iodide. 13. (See Basic radicals) a.
	4. Coloration, green. 7. Blue. 8. Coloration, deep-red.	3. Odour. 6. { Blackens paper moistened with acetate of lead.	
Not precipitated.	Not precipitated.		
1. **Tartrate,** white. sol in KHO. 2. **Citrate,** β. white. insol. in KHO.			
	1. **Succinate,** red-brown. 2. **Benzoate,** buff. 3. **Gallate,** bluish-black. 4. **Tannate,** bluish-black.		
	1. **Coloration,** deep red.		
a These are precipitated from a neutral solution. β Only precipitated on boiling.			a These acids are converted by H_2S into their corresponding sulphur acids, and are detected and separated in the process for basic radicals.

METHOD OF ANALYSIS FOR GROUP I.

(INORGANIC AND ORGANIC ACIDS.)

CHLORIDE OF BARIUM

precipitates from a *neutral* solution,

Sulphuric acid,	Phosphoric acid,	Boracic acid,
Hydrofluosilicic acid,	Oxalic acid,	Tartaric acid,
(Silicic acid),	Hydrofluoric acid,	Citric acid.

The other members of this Group (page 42) will, if present, have been detected during the course of analysis for basic radicals, or during the preparation of the solution before examining for acid radicals.

SILICIC ACID will be precipitated, if not previously detected and removed.

Care must be exercised to have the solution perfectly *neutral* before adding chloride of barium. If the solution was originally alkaline, it is most effectually rendered neutral by evaporating until the solution shews no alkaline reaction with litmus-paper. If acid, hydrate of ammonium is employed to neutralize.

BORATE of Barium, TARTRATE of Barium, and CITRATE of Barium are soluble in salts of ammonium. If the solution, previously acid, has been neutralized by hydrate of ammonium, the non-formation of a precipitate with chloride of barium will not indicate positively the absence of Boracic acid, Tartaric acid, or Citric acid, in consequence of the possible formation of chloride of ammonium.

If chloride of barium produces a precipitate, the *presence* of SULPHURIC ACID, HYDROFLUOSILICIC ACID, or PHOSPHORIC ACID, can only be inferred with any certainty.

If chloride of barium fails to produce a precipitate, the *absence* of SULPHURIC ACID, HYDROFLUOSILICIC ACID, PHOSPHORIC ACID, OXALIC ACID, or HYDROFLUORIC ACID, can only be certainly inferred.

ANALYSIS OF PRECIPITATE PRODUCED BY CHLORIDE OF BARIUM IN A *NEUTRAL* SOLUTION.

The precipitate is collected, washed, treated with hydrochloric acid in slight excess, and boiled gently.

Sulphate of barium and Fluosilicate of barium are insoluble in hydrochloric acid; the barium salts of the other acid radicals are soluble in this acid.

The insoluble residue may be rendered soluble in water by fusion with carbonate of sodium and nitrate of potassium. Acetic acid is added to the aqueous solution, and then chloride of calcium.

Sulphuric acid. Sulphate of calcium is insoluble, but Fluosilicate of calcium is soluble, in acetic acid. This insolubility indicates the presence of the Sulphuric radical.

Hydrofluosilicic acid. The presence or absence of the Fluosilicic radical in the acetic acid solution may be determined by the "etching test" (page 44, III.).

The solution is neutral.
Add $BaCl$.
1............Filter.
Wash ppt.
Add HCl to *ppt.*
Boil gently.
2............Filter.
(Examine filtrate as on next page.)

Residue from 2.
Fuse with $Na_2CO_3 + KNO_3$.
Digest with H_2O.
Add $HC_2H_3O_2$.
Add CaCl.
3............Filter.
Insoluble residue
indicates
presence of
Sulphuric acid.

Filtrate from 3.
Heat with conc[d]. H_2SO_4.
Dense fumes
which etch glass
indicate the presence of
Hydrofluosilicic acid.

METHOD OF ANALYSIS FOR GROUP I. (CONTINUED).

The filtered hydrochloric acid solution is treated with hydrate of ammonium.

Phosphate, Oxalate, and Fluoride of barium may be re-precipitated.

Borate, Oxalate, Fluoride, Tartrate and Citrate of barium, and possibly Silicate of barium, may be contained in the solution.

(If no re-precipitation occurs, the absence of Phosphoric acid can alone be affirmed with any certainty.)

The *precipitate* is re-dissolved in hydrochloric acid, boiled with carbonate of sodium, to convert into sodium salts (if necessary filtering from any precipitate of carbonate of barium), and re-precipitated as calcium salts by adding chloride of calcium.

Phosphoric acid.

Phosphate of calcium is soluble in acetic acid. After adding acetic acid the solution may be tested for the Phosphoric radical by ferric chloride or molybdate of ammonium (page 46).

Oxalate or Fluoride of calcium may be thrown down along with the Phosphate. They are insoluble in acetic acid.

Oxalic acid.

One portion of the insoluble residue may be tested for Fluorine by concentrated sulphuric acid (page 47).

Hydrofluoric acid.

Another portion, for the Oxalic radical, by concentrated sulphuric acid and binoxide of manganese (page 47).

The *filtrate* from the hydrate of ammonium precipitate, or the solution in which that reagent has failed to produce a precipitate, may contain Boracic acid, Oxalic acid, Hydrofluoric acid, Tartaric acid, Citric acid, or Silicic acid.

The solution is divided into two portions.

(Silicic acid.)

A. This portion is evaporated to dryness, treated with hydrochloric acid and water, and again evaporated, in order to convert any Silicic acid into insoluble silicic anhydride.

The solution which remains after this operation is divided into three portions.

Boracic acid.

a. Test for the Boracic radical with turmeric paper (page 46, note *a*).

Hydrofluoric acid.

b. Test for Fluorine, by means of concentrated sulphuric acid and sand, or by the "etching test" (page 47).

Oxalic acid.

c. Test for Oxalic acid, with concentrated sulphuric acid, and binoxide of manganese (page 47).

Citric acid.

B. This portion of the solution is treated with a small quantity of hydrate of potassium, in order to decompose any chloride of ammonium which may have been formed (page 55, notes *a*). Chloride of calcium is then added, and the whole boiled. After allowing to cool, more hydrate of potassium is added. A precipitate indicates the presence of the Citric radical, since citrate of calcium is insoluble, whilst Tartrate of calcium is soluble in hydrate of potassium.

Tartaric acid.

Test the filtered solution for the Tartaric radical, with a crystal of nitrate of silver (page 55, note *a*).

Filtrate from 2, preceding page.
Add NH_4HO.
1............Filter.

Residue from 1.
Dissolve in HCl.
Boil with Na_2CO_3.
(Filter from any ppt.)
Add CaCl.
Add $HC_2H_3O_2$.
2............Filter.
Add Fe_2Cl_6 to *filtrate*.
White-yellow ppt. $=(Fe_2)_3PO_4$?
or test with Molyb. Am.

Residue from 2.
Divide into two portions.
A. Heat with conc^d. H_2SO_4.
Dense fumes = HF.
Fumes etch glass.
B. Heat with conc^d. $H_2SO_4 + Mn_2O_2$.
Effervescence indicates presence of **Oxalic acid.**

Filtrate from 1, or solution after adding NH_4HO.
Divide into two portions.
A. Evaporate to dryness.
Add HCl and H_2O.
Evaporate again.
3............Filter.
Residue = Si_2O_3.
Divide solution from 3. into three portions.

a. Test with turmeric paper.
Red coloration indicates presence of **Boracic acid.**

b. Heat with conc^d. H_2SO_4 + sand. (p. 47, VI. ii.)
Deposit of Silica indicates presence of **Hydrofluoric acid.**

c. Heat with conc^d. $H_2SO_4 + Mn_2O_2$.
Effervescence indicates presence of **Oxalic acid.**

B. Add a little KHO.
Add CaCl.
Boil.
When cool, add KHO.
White ppt. $= Ca_3C_6H_5O_7$.
Citric acid.
4............Filter.
Add NH_4HO *to filtrate*.
Add crystal of $AgNO_3$.
Heat gently.
Metallic silver indicates presence of **Tartaric acid.**

METHOD OF ANALYSIS FOR GROUP II.

NITRATE OF SILVER

precipitates from an *acid* solution (acidified with **nitric** acid),

Hydrochloric acid,	Hydriodic acid,
Hydrobromic acid,	Hydroferrocyanic acid,
Hydroferricyanic acid,	Hydrosulphocyanic acid,

Hydrocyanic acid.

Nitrate of silver also precipitates many other acid radicals, if the solution is not decidedly acid.

If the addition of nitric acid before nitrate of silver produces any precipitate, this will be due to the presence of certain acid radicals of the first Group. The precipitate may be filtered off and neglected.

ANALYSIS OF PRECIPITATE PRODUCED BY NITRATE OF SILVER IN AN ACID SOLUTION.

The *precipitate* is treated with dilute nitric acid, to ensure the removal of all acids except those belonging to this group. After filtering, the precipitate is treated with hydrate of ammonium and heated. This operation serves to separate the group into two sections.

The *residue* may contain Hydriodic acid, Hydroferrocyanic acid, or Hydrosulphocyanic acid.

The *solution* may contain Hydrochloric acid, Hydrobromic acid, and possibly Hydrocyanic acid.

Hydriodic acid. A portion of the *residue* is heated with concentrated nitric acid. Hydriodic acid will manifest its presence by the violet vapours of iodine which accompany the decomposition.

(Iodic acid.) [If Hydriodic acid is detected, it will be necessary to test a portion of the original solution with a view to determine whether iodine existed in the form of iodide or iodate. They are distinguished by the difference in the reactions with concentrated sulphuric acid, since an *iodate* does not evolve free iodine when heated with this acid (pages 49 and 50).]

Hydrosulphocyanic acid. Hydrosulphocyanic acid yields a *yellow precipitate* when heated with concentrated nitric acid.

Hydroferrocyanic acid. The remaining portion of the *residue* is boiled with hydrate or carbonate of potassium or sodium, to convert any ferrocyanogen into **soluble ferrocyanide** of potassium or sodium. The radical may then **be sought for** according to the plan given in note γ, page 53, **or with ferric** chloride (in the latter case previously adding some **hydrochloric acid**).

The solution is rendered *acid*, by adding HNO_3.
(Filter any ppt. and neglect it.)
Add $AgNO_3$.
1............Filter.

Residue from 1.
Add dilute HNO_3.
2............Filter.
Add NH_4HO *to ppt.*
3............Filter.
(Examine filtrate as on next page.)

Divide *residue* into two portions.

A. Heat with concd. HNO_3.
Violet vapours = I.
(Examine original solution for Hydriodic and Iodic acid.)
Yellow ppt.
indicates presence of
Hydrosulphocyanic acid.

B. Add KHO or NaHO
or
K_2CO_3 or Na_2CO_3.
Boil.
Add to solution HCl.
Add Fe_2Cl_6.
Blue ppt. = $(Fe_2)_2(Cfy)_3$
or
test as in note γ, page 53
for
Hydroferrocyanic acid.

METHOD OF ANALYSIS FOR GROUP II. (CONTINUED).

The *solution* (after adding hydrate of ammonium to the nitrate of silver precipitate) may contain Hydrochloric acid, Hydrobromic acid, and possibly Hydrocyanic acid.

(Hydrocyanic acid will probably have been converted into a sulphocyanide, and ferricyanogen into ferrocyanogen by the action of hydrosulphuric acid, in preparing the solution.)

The solution is treated with excess of nitric acid to re-precipitate as silver salts.

The *precipitate* (after washing by decantation) is ignited in a crucible, in order to decompose the Cyanogen compounds of silver, the silver being thereby reduced to the metallic state. The odour during this operation will probably lead to the detection of Cyanogen, if it is present.

The fused mass is treated with boiling dilute nitric acid.

The solution will contain nitrate of silver, and if now hydrochloric acid is added, a precipitate will indicate the previous presence of Hydrocyanic acid or Hydroferricyanic acid.

Any residue, after fusion, may contain the radicals Chlorine or Bromine.

This residue is fused again with carbonate of potassium and sodium, and dissolved in water.

One portion is treated for Bromine by means of sulphuric acid and starch-paste (p. 50, IV. i.).

If Bromine is present, another portion is tested for Chlorine, with sulphuric acid and chromate of potassium (p. 50, IV. ii.). If this latter course becomes necessary, it will however be advisable to operate with a portion of the original solution.

If Bromine is absent, the residue can only contain Chlorine, and further examination becomes unnecessary.

A special examination of a portion of the *original solution* must be undertaken, if the presence of Hydrocyanic acid on Hydroferricyanic acid is suspected. The former may be detected by the reactions with sulphide of ammonium (p. 51); the latter by the negative results when testing for Hydrocyanic acid with sulphide of ammonium, and the reactions on subsequently adding ferrous sulphate.

Filtrate from 3, preceding page.
Add excess of HNO_3.
1............Filter.
Wash ppt. by decantation.
Ignite the ppt.
(Observe if any odour.)
Add to fused mass HNO_3.
2............Filter.
Add HCl *to filtrate.*
White ppt. = AgCl
indicating presence of Cyanide *or* Ferricyanide of silver in previous ppt.

Residue from 2.
Fuse with $Na_2CO_3 + K_2CO_3$.
Dissolve in H_2O.
Divide into two portions.
A. Add concd. H_2SO_4 and some starch-paste.
Heat.
Starch-paste coloured yellow
indicates presence of
Bromine.
B. *If Bromine is present.*
Evaporate (or portion of the original solution) to dryness.
Heat with concd. $H_2SO_4 + KCrO_4$.
Condense vapours.
Add NH_4HO to liquid.
Yellow colour
indicates presence of
Chlorine.
Acetic acid gives a *crimson ppt.* with this yellow solution.

Examine specially for Hydrocyanic acid and Hydroferricyanic acid in *original solution.*
Test as in page 51, III. for Hydrocyanic acid.
Test as in p. 50, II. for Hydroferricyanic acid.

METHOD OF ANALYSIS FOR GROUP III.

Nitric acid,
and
Chloric acid,
are *not precipitated* by the reagents employed to precipitate the former groups.

SULPHINDIGOTIC ACID

produces a yellow colour in a solution which contains one or both of these acids.

Nitric acid. If this coloration with a solution of indigo has taken place, the presence of Nitric acid is revealed by the characteristic reactions with ferrous sulphate in presence of sulphuric acid (p. 54, I.).

Chloric acid. Chloric acid is identified by the decomposition effected by sulphurous acid, and the subsequent reaction with nitrate of silver (p. 54, IV.).

METHOD OF ANALYSIS FOR GROUP II.

(ORGANIC ACIDS.)

FERRIC CHLORIDE

precipitates from a *neutral* solution,

Succinic acid, red-brown **Benzoic acid,** buff.		Benzoate of iron dissolves in hydrochloric acid, with separation of crystalline Benzoic acid. Distinguished also by reactions with chloride of barium and alcohol (p. 56).
Gallic acid, **Tannic acid,**	bluish-black.	Distinguished by reactions with ferric chloride, or solution of gelatine (p. 57).

produces *coloration* in a solution in presence of

Acetic acid, Hydrosulphocyanic acid,	deep red colour.	Colour disappears, if Acetic acid only is present, on adding mercuric chloride, or on boiling. Test also for Acetic acid with sulphuric acid and alcohol (p. 58).
Hydroferricyanic acid.	green tint.	

[Phosphoric acid (white) and Hydroferrocyanic acid (blue) may also be precipitated from a neutral solution by ferric chloride].

ACIDS WHICH ARE DETECTED DURING THE ANALYSIS FOR BASIC RADICALS, OR DURING THE PREPARATION OF THE SOLUTION TO BE EXAMINED FOR ACID RADICALS.

Carbonic Acid, Hydrosulphuric Acid, Hydrocyanic Acid, are detected by the effervescence, on adding the group-reagent, hydrochloric acid. The characteristic reactions by which they are identified, are given on p. 85. They are invariably detected when examining for basic radicals (p. 64).

Sulphurous Acid, Hyposulphurous Acid, are decomposed, with separation of sulphur, by hydrosulphuric acid, if hydrochloric acid is present. Hydrochloric acid causes a deposit of sulphur in the case of Hyposulphurous acid, but not in the case of Sulphurous acid. The latter is detected in the preliminary examination (p. 82).

APPENDIX.

ANALYSIS OF SUBSTANCES, INSOLUBLE IN WATER AND IN ACIDS.

(FRESENIUS.)

All, otherwise insoluble, bodies which come under the student's notice, in an elementary course, will, after fusion with alkaline carbonate (page 21, par. 8), yield to the solvent action of water, hydrochloric acid, or nitric acid.

The following method, adapted from Fresenius, indicates the mode of dealing with such bodies in the course of actual analysis.

The substances to which this method is applicable are the following:

- Barium, Strontium, Lead, as Sulphates.
- Silver, as Chloride, Iodide, Bromide, and Cyanide.
- Lead, as Chloride.
- Calcium, as Fluoride (and certain other Fluorides).
- Silicates.
- Aluminium, chiefly as Sesquioxide.
- Chromium, as Sesquioxide.
- Certain Salts of Tin, Antimony, and Arsenic.
- (Sulphur and Carbonaceous matter.)

TREATMENT OF INSOLUBLE SUBSTANCES, AFTER MAKING A PRELIMINARY EXAMINATION OF THE SUBSTANCE UNDER ANALYSIS.

Heat for some time with acetate of ammonium (conc[d]).

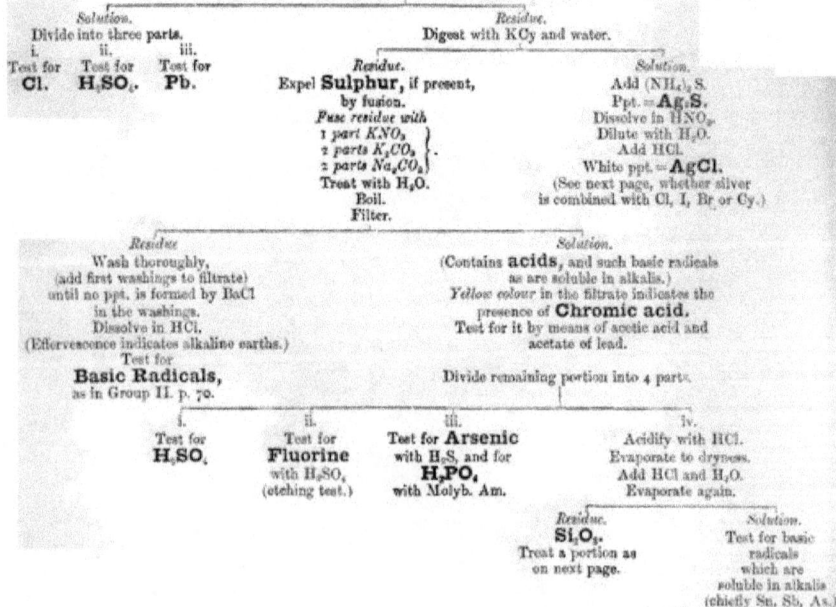

- *Solution.* Divide into three parts.
  - i. Test for **Cl.**
  - ii. Test for **H_2SO_4.**
  - iii. Test for **Pb.**
- *Residue.* Digest with KCy and water.
  - *Residue.* Expel **Sulphur,** if present, by fusion. *Fuse residue with* *1 part KNO_3, 2 parts K_2CO_3, 2 parts Na_2CO_3*. Treat with H_2O. Boil. Filter.
    - *Residue.* Wash thoroughly, (add first washings to filtrate) until no ppt. is formed by BaCl in the washings. Dissolve in HCl. (Effervescence indicates alkaline earths.) Test for **Basic Radicals,** as in Group II. p. 70.
    - *Solution.* (Contains **acids,** and such basic radicals as are soluble in alkalis.) *Yellow colour* in the filtrate indicates the presence of **Chromic acid.** Test for it by means of acetic acid and acetate of lead.

      Divide remaining portion into 4 parts.
      - i. Test for **H_2SO_4**
      - ii. Test for **Fluorine** with H_2SO_4 (etching test.)
      - iii. Test for **Arsenic** with H_2S, and for **H_3PO_4** with Molyb. Am.
      - iv. Acidify with HCl. Evaporate to dryness. Add HCl and H_2O. Evaporate again.
        - *Residue.* **Si_2O_3.** Treat a portion as on next page.
        - *Solution.* Test for basic radicals which are soluble in alkalis (chiefly Sn, Sb, As.)
  - *Solution.* Add $(NH_4)_2S$. Ppt. = **Ag_2S.** Dissolve in HNO_3. Dilute with H_2O. Add HCl. White ppt. = **AgCl.** (See next page, whether silver is combined with Cl, I, Br or Cy.)

ANALYSIS OF INSOLUBLE SUBSTANCES (CONTINUED.)

ANALYSIS OF ALKALINE SILICATES.

The insoluble substance which may contain an alkali in combination with the silicic radical, is decomposed by fusion with four parts of hydrate of barium, for a considerable time at a high temperature.

The fused mass is then treated as follows

Treat with hydrochloric acid and water, to dissolve.
Treat the solution with hydrate and carbonate of ammonium, to precipitate the Barium.
Filter.
Evaporate filtrate to dryness.
Ignite.
Dissolve the residue in water.
Precipitate again with hydrate and carbonate of ammonium.
Filter.
Evaporate to dryness.
Ignite to expel ammonia.
Test the residue for **Potassium** and **Sodium** in the usual manner.

TO ASCERTAIN WHETHER THE INSOLUBLE SALT OF SILVER IS A CHLORIDE, IODIDE, BROMIDE, OR CYANIDE.

Treat a portion of the *original substance* with water to remove soluble matter.
And then with nitric acid for the same purpose.
Wash the residue with water.
Boil with rather dilute solution of hydrate of sodium.
Filter. (1)
Acidify *the filtrate* with nitric acid, and test for
Ferrocyanogen and **Ferricyanogen.**
(This may be effected by means of ferric chloride.)
Digest the residue from (1) with granulated zinc and water.
Add some dilute sulphuric acid.
(The object of this is to reduce the silver to the metallic state.)
Filter, in ten minutes.
Add carbonate of sodium *to filtrate*, to precipitate zinc.
Test filtrate specially for
Chlorine, Bromine, Iodine, and **Cyanogen.**

ANALYSIS OF ALLOYS.

The principle of analysis depends on the fact that nitric acid converts certain metals into nitrates, in which condition they are soluble, whilst nitric acid either fails to attack other metals, or merely oxidising them fails to reduce them to a soluble form.

The alloy must be reduced to as complete a state of subdivision as is possible.

Boil gently with tolerably concentrated nitric acid.
Add a considerable quantity of water.
Boil to remove excess of acid.
Filter.

Residue may contain GOLD, PLATINUM, ANTIMONY, TIN.
(Small traces of *Antimony* may be completely dissolved by nitric acid. Small quantities of *Platinum* alloyed with *Silver* may also dissolve in nitric acid.)

If the residue is *white*, it probably contains TIN, or ANTIMONY.

If the residue is *metallic or a black mass*, it probably contains, GOLD, or PLATINUM.

The residue is well washed.
Boil with nitro-hydrochloric acid, (three parts hydrochloric to one part of nitric acid).

- *Residue* will be a white insoluble powder. Consists of CHLORIDE OF SILVER or possibly CHLORIDE OF LEAD. Treat as a substance insoluble in water and acids, p. 92.
- *Solution.* Divide the solution into two parts.
  - i. Test for GOLD and PLATINUM, as on p. 70.
  - ii. Test for ANTIMONY and TIN, as on p. 72.

Solution. Analyse this, in the usual manner, for the metals.
SILVER.
LEAD.
MERCURY.
ARSENIC.
BISMUTH.
COPPER.
IRON.
ALUMINIUM.
ZINC.
COBALT.
NICKEL.

www.ingramcontent.com/pod-product-compliance
Lightning Source LLC
Chambersburg PA
CBHW021944190326
41519CB00009B/1139
9783337279868